GRADE 2

STAAR Mathematics

PRACTICE

Table of Contents

Using This Book

What Is the STAAR Mathematics Assessment?

The State of Texas Assessments of Academic Readiness (STAAR) is the current assessment for students in the state of Texas. STAAR Mathematics assesses what students are expected to learn at each grade level according to the developmentally appropriate academic readiness and supporting standards outlined in the Texas Essential Knowledge and Skills (TEKS).

How Does This Book Help My Student(s)?

If your student is taking the STAAR Assessment for Mathematics, then as a teacher and/or parent you can use the mini-lessons, math practice pages, and practice tests in this book to prepare for the STAAR Mathematics exam. This book is appropriate for on-grade-level students.

STAAR Mathematics Practice provides:

- Mini-lessons for assessed Math TEKS skills and strategies
- Word problems for assessed Math TEKS skills and strategies
- Questions for griddable and multiple-choice answer format
- Opportunities to familiarize students with STAAR format and question stems
- Answer Keys available online for access anywhere

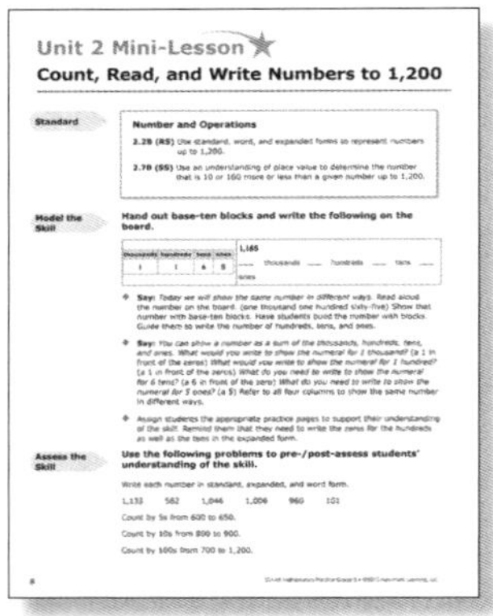

Introduce STAAR-aligned math concept, skill, or strategy

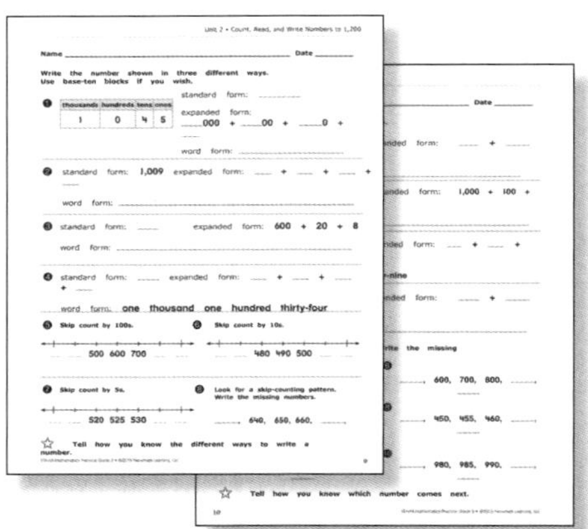

Practice with STAAR-aligned problems

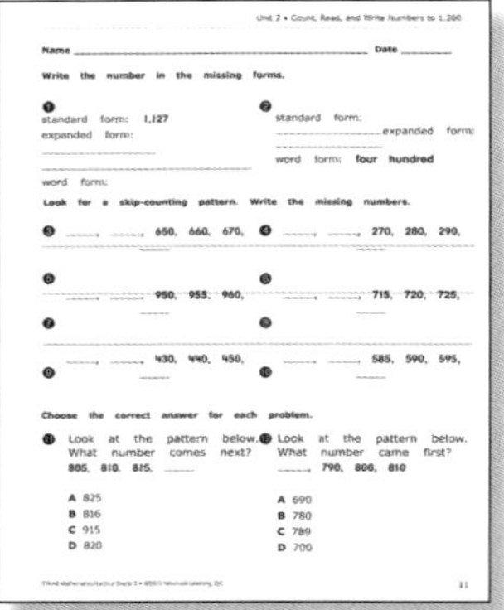

Assess concepts, skills, and strategies with word problems

STAAR Mathematics Practice/TEKS Alignment Chart • Grade 2*

(*Currently, there is no STAAR Mathematics Assessment for Grade 2, but these skills are a foundation for Grade 3 TEKS later assessed.)

STAAR Mathematics Practice Unit	2.1	2.2	2.3	2.4	2.5	2.6	2.7	2.8	2.9	2.10	2.11
Unit 1: Understand Place Value		✔									
Unit 2: Count, Read, and Write Numbers to 1,200		✔					✔				
Unit 3: Compare Numbers		✔									
Unit 4: Order Numbers	✔	✔									
Unit 5: Fractions of a Whole	✔		✔								
Unit 6: Fractions of a Group	✔		✔								
Unit 7: Use Models to Compare Fractions	✔		✔								
Unit 8: Addition and Subtraction Fact Families				✔							
Unit 9: Use Strategies to Add				✔							
Unit 10: Add Two-Digit Numbers				✔							
Unit 11: Use Strategies to Subtract				✔							
Unit 12: Subtract Two-Digit Numbers				✔							
Unit 13: Solve Word Problems	✔			✔							
Unit 14: How Much Money?					✔						
Unit 15: Model Multiplication						✔					
Unit 16: Model Division						✔					
Unit 17: Find Patterns	✔										
Unit 18: Make a Table	✔										
Unit 19: Extend Patterns	✔										
Unit 20: Describe Plane Figures								✔			
Unit 21: Describe Solid Figures								✔			
Unit 22: Create New Figures								✔			
Unit 23: Locate Points on a Number Line		✔									
Unit 24: Use Models for Inch and Foot	✔								✔		
Unit 25: Use Models for Centimeter and Meter	✔								✔		
Unit 26: Use Non-Standard Units to Determine Area									✔		
Unit 27: Use Non-Standard Units to Determine Capacity and Weight	✔										
Unit 28: Read a Thermometer	✔										
Unit 29: Tell Time to the Nearest Minute									✔		
Unit 30: Estimate How Much Time It Takes	✔										
Unit 31: Make a Line Plot										✔	
Unit 32: Make a Graph										✔	
Unit 33: More Likely or Less Likely	✔										
Unit 34: How Much Money Is Saved?											✔
Unit 35: Deposits and Withdrawals											✔

Understand Place Value

Standard

Number and Operations

2.2A (SS) - Use concrete and pictorial models to compose and decompose numbers up to 1,200 in more than one way as a sum of so many thousands, hundreds, tens, and ones.

Model the Skill

Hand out base-ten blocks.

- **Say:** *A number can be shown with base-ten blocks.* Review with students that 10 ones equal 1 ten. Then show 10 tens. **Ask:** *What number is shown with 10 tens?* (100) Have students count the rods by tens up to 100. **Ask:** *What number can be shown with 10 hundreds?* (1,000) Have the students count the squares by hundreds up to 1,000. **Say:** *Use blocks to show the number 1,123. How many thousands do you use?* (1) *How many tens?* (2) *How many ones?* (3)
- **Ask:** *How do you show the number 200?* (Possible answer: with 2 hundreds) *Why are there no tens or ones shown in the chart?* Write the number 300 on the board. **Ask:** *Why are there zeros in the tens and ones places?* (Possible answer: because there are no tens or ones in the number 300) *Even though only 3 hundreds are needed to show the number, you need to write the zeros in the tens and ones places to indicate that no tens or ones are needed.*
- Assign students the appropriate practice page(s) to support their understanding of the skill.

Assess the Skill

Use the following problem to pre-/post-assess students' understanding of the skill.

Ask students to use the place value chart to describe the number of thousands, hundreds, tens, and ones for each value.

	thousands	hundreds	tens	ones
1,147				
981				
1,052				

Name ________________________________ Date __________

Write the number in the correct places of the chart.

1

1,153

thousands	hundreds	tens	ones

2

1,060

thousands	hundreds	tens	ones

3

308

thousands	hundreds	tens	ones

4

1,179

thousands	hundreds	tens	ones

5

1,200

thousands	hundreds	tens	ones

6

1,003

thousands	hundreds	tens	ones

7

469

thousands	hundreds	tens	ones

 **Tell how you know what number to write in each place.**

Name ________________________________ Date __________

Write the number of thousands, hundreds, tens, and ones. Then write the number.

1

thousands	hundreds	tens	ones
	3	1	8

____ thousands
____ hundreds
____ tens
____ ones = ________

2

thousands	hundreds	tens	ones
1	0	0	2

____ thousands
____ hundreds
____ tens
____ ones = ________

3

thousands	hundreds	tens	ones
1	0	9	1

____ thousands
____ hundreds
____ tens
____ ones = ________

4

thousands	hundreds	tens	ones
	6	4	2

____ thousands
____ hundreds
____ tens
____ ones = ________

5

thousands	hundreds	tens	ones
	4	0	9

____ thousands
____ hundreds
____ tens
____ ones = ________

6

thousands	hundreds	tens	ones
1	1	9	9

____ thousands
____ hundreds
____ tens
____ ones = ________

7

thousands	hundreds	tens	ones
1	1	4	0

____ thousands
____ hundreds
____ tens
____ ones = ________

8

thousands	hundreds	tens	ones
		7	7

____ thousands
____ hundreds
____ tens
____ ones = ________

9

thousands	hundreds	tens	ones
1	0	2	8

____ thousands
____ hundreds
____ tens
____ ones = ________

10

thousands	hundreds	tens	ones
	7	0	0

____ thousands
____ hundreds
____ tens
____ ones = ________

11

thousands	hundreds	tens	ones
	2	5	0

____ thousands
____ hundreds
____ tens
____ ones = ________

12

thousands	hundreds	tens	ones
1	1	5	0

____ thousands
____ hundreds
____ tens
____ ones = ________

Tell how you know the number of tens.

Name ______________________ **Date** __________

Write the number of thousands, hundreds, tens, and ones.

1. The number 1,114 has —

_______ thousands
_______ hundreds
_______ tens
_______ ones

2. The number 4 has —

_______ thousands
_______ hundreds
_______ tens
_______ ones

3. The number 1,064 has —

_______ thousands
_______ hundreds
_______ tens
_______ ones

4. The number 53 has —

_______ thousands
_______ hundreds
_______ tens
_______ ones

5. The number 442 has —

A Four ones, four hundreds, and two tens
B Four hundreds, two tens, and two ones
C Four hundreds, four tens, and ten ones
D Four hundreds, four tens, and two ones

6. The number 1,078 has —

A One thousands, zero hundreds, seven ones, and eight tens
B One thousands, one hundreds, seven tens, and eight ones
C One thousands, zero hundreds, seven tens, and eight ones
D One hundreds, zero thousands, seven tens, and eight ones

Unit 2 Mini-Lesson

Count, Read, and Write Numbers to 1,200

Standard

Number and Operations

2.2B (RS) Use standard, word, and expanded forms to represent numbers up to 1,200.

2.7B (SS) Use an understanding of place value to determine the number that is 10 or 100 more or less than a given number up to 1,200.

Model the Skill

Hand out base-ten blocks and write the following on the board.

thousands	hundreds	tens	ones
1	1	6	5

1,165

____ thousands ____ hundreds ____ tens ____ ones

Expanded Form: ____000 + ____00 + ____0 + ______

- **Say:** *Today we will show the same number in different ways.* Read aloud the number on the board. (one thousand one hundred sixty-five) Show that number with base-ten blocks. Have students build the number with blocks. Guide them to write the number of hundreds, tens, and ones.

- **Say:** *You can show a number as a sum of the thousands, hundreds, tens, and ones. What would you write to show the numeral for 1 thousand?* (a 1 in front of the zeros) *What would you write to show the numeral for 1 hundred?* (a 1 in front of the zeros) *What do you need to write to show the numeral for 6 tens?* (a 6 in front of the zero) *What do you need to write to show the numeral for 5 ones?* (a 5) Refer to all four columns to show the same number in different ways.

- Assign students the appropriate practice pages to support their understanding of the skill. Remind them that they need to write the zeros for the hundreds as well as the tens in the expanded form.

Assess the Skill

Use the following problems to pre-/post-assess students' understanding of the skill.

Write each number in standard, expanded, and word form.

1,133 582 1,046 1,006 960 101

Count by 5s from 600 to 650.

Count by 10s from 800 to 900.

Count by 100s from 700 to 1,200.

Name ______________________________ **Date** __________

Write the number shown in three different ways.
Use base-ten blocks if you wish.

1.

thousands	hundreds	tens	ones
1	0	4	5

standard form: __________

expanded form:
____**000** + ____**00** + ____**0** + ____

word form: ______________________________
____

2. standard form: **1,009** expanded form: ____ + ____ + ____ + ____

word form: ______________________________

3. standard form: _____ expanded form: **600 + 20 + 8**

word form: ______________________________

4. standard form: _____ expanded form: ____ + ____ + ____ + ____

word form: **one thousand one hundred thirty-four**

5. **Skip count by 100s.**

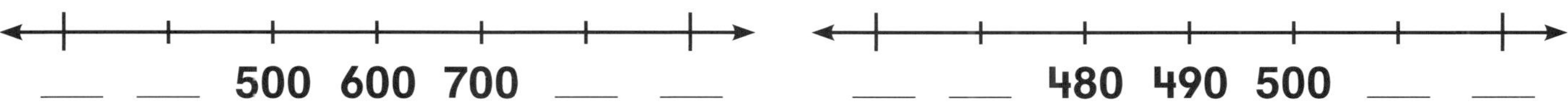

6. **Skip count by 10s.**

7. **Skip count by 5s.**

8. **Look for a skip-counting pattern.**
Write the missing numbers.

_____ 640 650 660 _____ _____

 **Tell how you know the different ways to write a number.**

Name ______________________________ **Date** __________

Fill in the blanks for each problem.

1. standard form: **732** expanded form: _____ + _____ + _____

   word form: ______________________________

2. standard form: _____ expanded form: **1,000 + 100 + 90 + 4**

   word form: ______________________________

   ____

3. standard form: _____ expanded form: _____ + _____ + _____ + _____

   word form: **one thousand fifty-nine**

4. standard form: **813** expanded form: _____ + _____ + _____

   word form: ______________________________

Look for a skip-counting pattern. Write the missing numbers.

5. _____, **330, 335, 340,** _____, _____

6. _____, **600, 700, 800,** _____, _____

7. _____, **750, 760, 770,** _____, _____

8. _____, **450, 455, 460,** _____, _____

9. _____, **550, 650, 750,** _____, _____

10. _____, **980, 985, 990,** _____, _____

☆ **Tell how you know which number comes next.**

Name ______________________________ **Date** ________

Write the number in the missing forms.

1. standard form: **1,127**
expanded form: ______________________
____
word form: ______________________
____

2. standard form: ______________________
expanded form: ______________________
word form: **four hundred seventy**

Look for a skip-counting pattern. Write the missing numbers.

3. ____, ____, **650, 660, 670,** ____

4. ____, ____, **270, 280, 290,** ____

5. ____, ____, **950, 955, 960,** ____

6. ____, ____, **715, 720, 725,** ____

7. ____, ____, **430, 440, 450,** ____

8. ____, ____, **585, 590, 595,** ____

9. ____, ____, **835, 840, 845,** ____

10. ____, ____, **640, 650, 660,** ____

Choose the correct answer for each problem.

11. Look at the pattern below. What number comes next?
805, 810, 815, ____

A 825
B 816
C 915
D 820

12. Look at the pattern below. What number came first?
____, **790, 800, 810**

A 690
B 780
C 789
D 700

Unit 3 Mini-Lesson
Compare Numbers

Standard

Number and Operations

2.2D (RS) Use place value to compare and order whole numbers up to 1,200 using comparative language, numbers, and symbols (<,>, or =).

Model the Skill

Hand out base-ten blocks and write the following values on the board.

123 132

- **Say:** *Let's compare these two numbers. Use your base-ten blocks to make each number.* Show the numbers in separate locations. **Say:** *First, you need to compare the greatest place—the hundreds. Does one number have more hundreds than the other?* (no) *Compare the next place—the tens. Does one number have more tens than the other?* (Yes, 132 has more tens.) Point out that there is no need to compare the ones because students have already determined which number is greater. **Say:** *132 is greater than 123 and 123 is less than 132.* Then practice another example with students.

- **Say:** *Now compare 205 and 210. This group of 205 must be greater because it has more blocks. Am I right?* (no) *Why not?* (Possible answer: Having more blocks doesn't mean the number is greater; you have to look at the values of the blocks.) *1 ten is equal to 10 ones. If you showed 210 with 2 hundreds and 10 ones, there would be more blocks. 210 is greater than 205 and 205 is less than 210.*

- Assign students the appropriate activity page(s) to support their understanding of the skill.

Assess the Skill

Use the following problem to pre-/post-assess students' understanding of the skill.

Use place value to compare numbers. Write <, >, =.

354 ◯ 364 732 ◯ 732 1,023 ◯ 1,123

Name ______________________________ Date __________

Compare each set of numbers. Circle the true statement. Then write the correct sign for each problem.

>	**is greater than**
<	**is less than**
=	**is equal to**

1

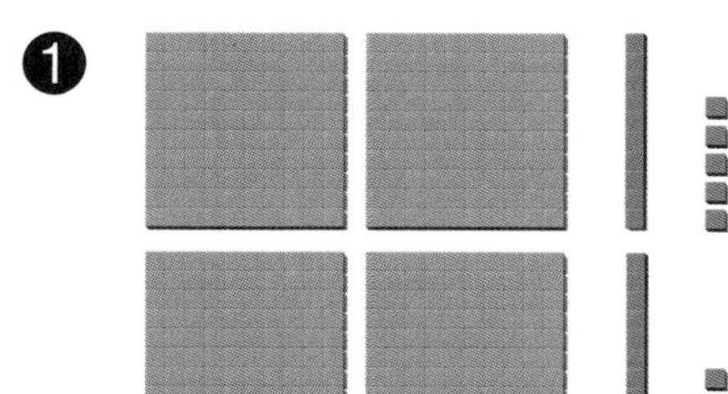

215 is greater than 213.
215 is less than 213.
215 is equal to 213.

215 ◯ 213

2

thousands	hundreds	tens	ones
1	1	9	5
1	0	3	7

1,195 is greater than 1,037.
1,195 is less than 1,037.
1,195 is equal to 1,037.

1,195 ◯ 1,037

3

hundreds	tens	ones
7	3	9
7	7	9

739 is greater than 779.
739 is less than 779.
739 is equal to 779.

739 ◯ 779

Write the correct sign for each problem.

4 347 ◯ 374

5 681 ◯ 918

6 1,120 ◯ 1,102

7 457 ◯ 475

8 1,167 ◯ 1,171

9 1,081 ◯ 1,081

 **Tell how you know when a number is less than the other number.**

Name ______________________________ **Date** __________

Compare each set of numbers.
Then write the correct sign for each problem.

>	**is greater than**
<	**is less than**
=	**is equal to**

1

hundreds	tens	ones
6	3	2
6	4	1

632 ◯ 641

2

thousands	hundreds	tens	ones
1	1	4	2
1	0	4	1

1,142 ◯ 1,041

3 631 ◯ 913

4 802 ◯ 802

5 654 ◯ 954

6 1,179 ◯ 1,181

7 218 ◯ 128

8 709 ◯ 706

9 454 ◯ 462

10 1,008 ◯ 1,011

11 1,177 ◯ 1,174

12 356 ◯ 356

Tell how you know which symbol to write.

Name ______________________________ **Date** __________

Solve.

1. Sally has 231 baseball cards. Matt has 229 baseball cards. Who has a greater number of baseball cards?

2. The cheese store sold 329 pounds of cheese on Monday. It sold 259 pounds of cheese on Tuesday. On which day did it sell more cheese?

Circle the correct answer for each problem.

3. Claudia has 1,123 beads in her collection. Jaden has 1,032 beads. Liv has 1,132 beads. Who has more beads?

   A Claudia
   B Liv
   C Jaden
   D They have an equal number of beads.

4. Riley has 439 baseball cards in her collection. If she trades 5 cards to Will for 5 cards, how many will Riley have?

   A 434 cards
   B 429 cards
   C 439 cards
   D 444 cards

Unit 4 Mini-Lesson
Order Numbers

Standard

Number and Operations

2.1A (PS) Apply mathematics to problems arising in everyday life, society, and the workplace.

2.2D (RS) Use place value to compare and order whole numbers up to 1,200 using comparative language, numbers, and symbols (<,>, or =).

Model the Skill

Hand out base-ten blocks and write the following values on the board.

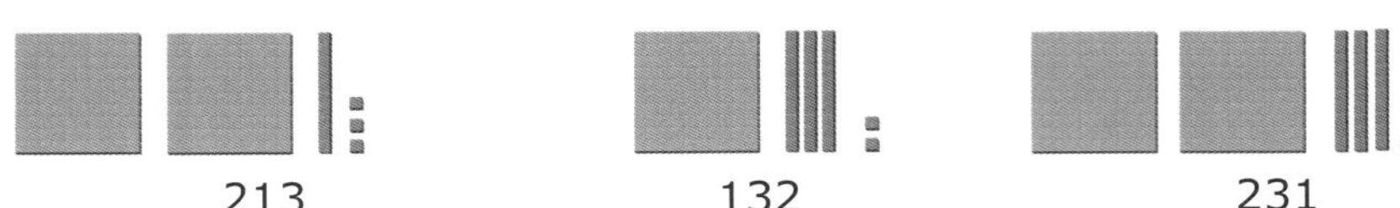

- **Say:** *Let's compare these three numbers. Use your base-ten blocks to make each number.* Show the numbers in separate locations. **Say:** *First, you need to compare the greatest place—the hundreds. Does one number have fewer hundreds than the others?* (Yes, 132 has fewer hundreds, so 132 has the least value of the three). *Compare the next place—the tens—for the remaining two numbers. Does one number have fewer tens than the other?* (Yes, 213 has fewer tens.) Point out that there is no need to compare the ones because students have already determined which numbers have the least, next to least, and greatest value. **Say:** *132 is less than 213 and 213 is less than 231.* Then practice another example with students.

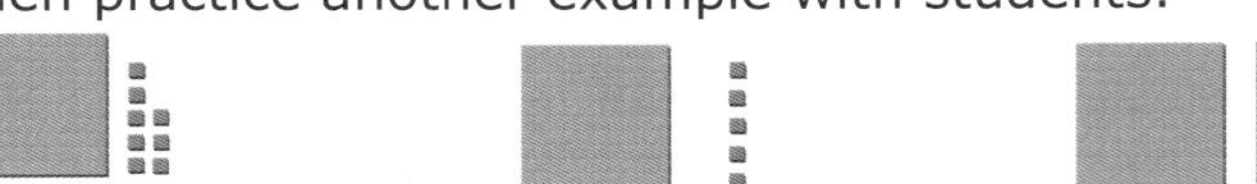

- **Say:** *Now compare 108, 105, and 110. This group of 108 must be greater because it has more blocks. Am I right?* (no) *Why not?* (Possible answer: Having more blocks doesn't mean the number is greater; you have to look at the values of the blocks.) *1 ten is equal to 10 ones. If you showed 110 with 1 hundred and 10 ones, there would be more blocks, but if you showed 110 with 1 hundred and 1 ten you would have the least number of blocks, but still the greatest value. 110 is greater than 105 and 108.*

- Assign students the appropriate activity page(s) to support their understanding of the skill.

Assess the Skill

Use the following problems to pre-/post-assess students' understanding of the skill.

Put these numbers in order of least to greatest in value.

402 1,039 413 699 960 331

Put these numbers in order of greatest to least in value.

732 739 537 808 1,077 1,127

Name ______________________________ Date __________

Compare each set of numbers.
Then write the numbers in order from least to greatest.

1

hundreds	tens	ones
6	3	2
6	4	1
6	3	4

_____ _____ _____

2

thousands	hundreds	tens	ones
1	0	2	0
1	2	0	0
1	0	0	2

_____ _____ _____ _____

3 631 913 739

_____ _____ _____

4 802 820 780

_____ _____ _____

5 654 954 695

_____ _____ _____

6 1,179 1,181 1,197

_____ _____ _____

7 218 128 188

_____ _____ _____

8 709 706 690

_____ _____ _____

9 1,054 1,145 1,045

_____ _____ _____

10 308 311 307

_____ _____ _____

11 577 574 575

_____ _____ _____

12 1,036 963 1,200

_____ _____ _____

☆ **Tell how you know when a number is less than the other number.**

Name ______________________ **Date** __________

Compare each set of numbers.
Then write the numbers in order from least to greatest.

1. 107 120 111

_____ _____ _____

2. 1,053 1,133 1,143

_____ _____ _____

3. 756 751 587

_____ _____ _____

4. 699 645 956

_____ _____ _____

5. 242 341 153

_____ _____ _____

6. 1,034 1,028 1,041

_____ _____ _____

7. 577 575 569

_____ _____ _____

8. 381 403 348

_____ _____ _____

9. 1,056 756 965

_____ _____ _____

10. 212 221 220

_____ _____ _____

11. 604 640 464

_____ _____ _____

12. 1,200 1,169 1,199

_____ _____ _____

☆ **Tell how you know which number goes last.**

Name ______________________________ **Date** __________

Solve.

1. Tanya is 37 inches tall. Finbar is 35 inches tall. Pilar is 38 inches tall. If they stood in height order, who would be standing in the middle?

2. The delicatessen store sold 1,013 sandwiches on Monday. It sold 1,103 sandwiches on Tuesday. It sold 1,031 sandwiches on Wednesday. On which day did it sell the least number of sandwiches?

Circle the correct answer for each problem.

3. Ruby's home is 572 meters from school. Nell's home is 575 meters from school. Jack lives 526 meters from school and Camilla lives 527 meters away from school. Who lives the farthest from school?

   A Ruby
   B Camilla
   C Nell
   D Jack

4. Bill, Sam, Alex, and Amelia were in a race. Bill ran to the finish in 236 seconds. Sam finished in 326 seconds, Alex finished in 263 seconds, and Amelia finished in 233 seconds. Who won the race?

   A Bill
   B Amelia
   C Alex
   D Sam

Unit 5 Mini-Lesson

Fractions of a Whole

Standard

Number and Operations

2.1A (PS) Apply mathematics to problems arising in everyday life, society, and the workplace.

2.3A (SS) Partition objects into equal parts and name the parts, including halves, fourths, and eighths, using words.

Model the Skill

Draw the following figure on the board.

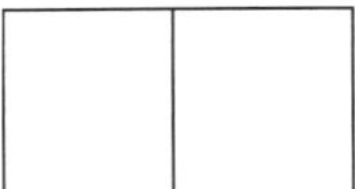

- **Say:** *Today we are going to learn about fractions. A fraction names part of a whole or part of a group. Look at this rectangle. How many equal parts is the rectangle divided into?* (2) *Each part is one-half of the whole rectangle.*

- Write the fraction $\frac{1}{2}$ on the board. Have students look at the fraction $\frac{1}{2}$ and point to the numerator. **Say:** *The numerator is the top number of a fraction. The numerator names a part of the whole.* Have students point to the denominator. **Say:** *The denominator is the bottom number of a fraction. The denominator tells how many equal parts are in the whole.*

- Model writing the fraction for the rectangle again. **Ask:** *How many equal parts are in the whole rectangle?* (2) *What fraction names each equal share or part?* ($\frac{1}{2}$) Have students practice writing the fraction for similar shapes sectioned in halves, thirds, and fourths or quarters.

- Assign students the appropriate practice page(s) to support their understanding of the skill and share their strategies for finding the missing numbers in each fraction. They should recognize that the denominator is always the same as the number of equal parts. Have them identify whether the numerator or the denominator is the missing number.

Assess the Skill

Use the following problems to pre-/post-assess students' understanding of the skill.

- **Ask:** *How many equal parts are in the whole? How many equal parts are shaded? What fraction shows the shaded part?*

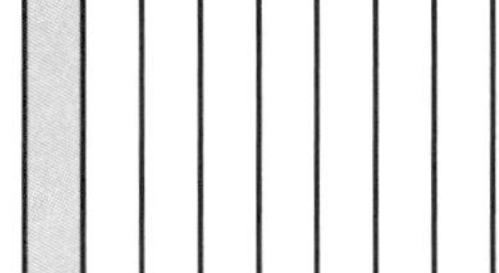 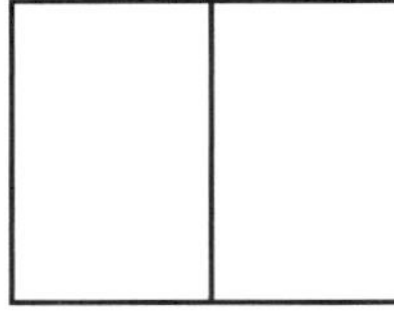 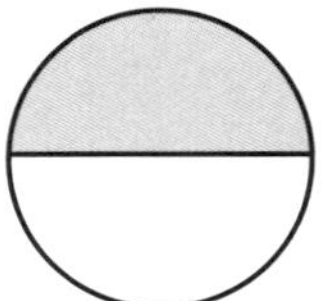 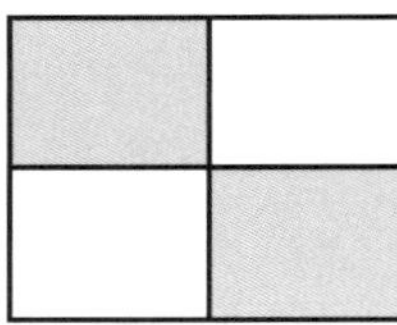 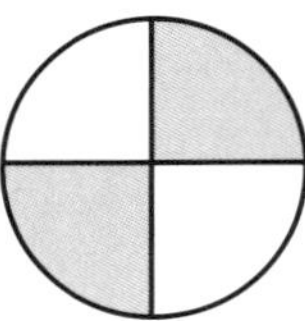 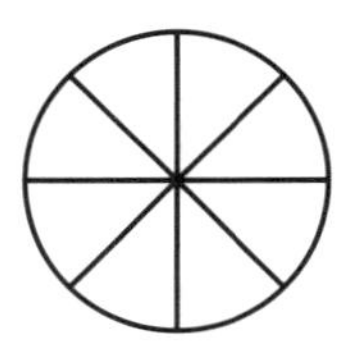

Name ______________________________ Date __________

Write each missing number.

1.

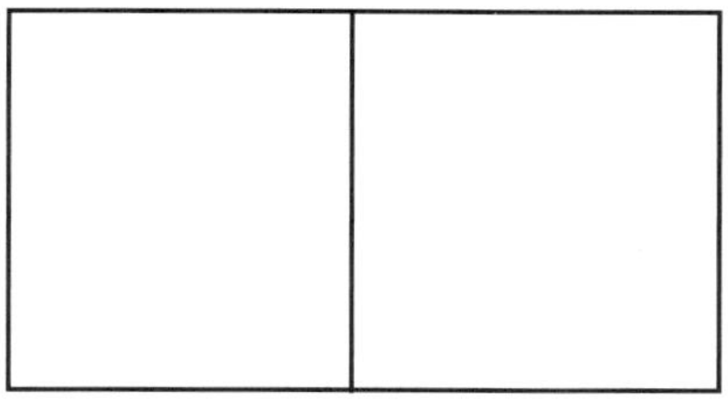

_______ equal parts

Each part is $\frac{1}{2}$.

2.

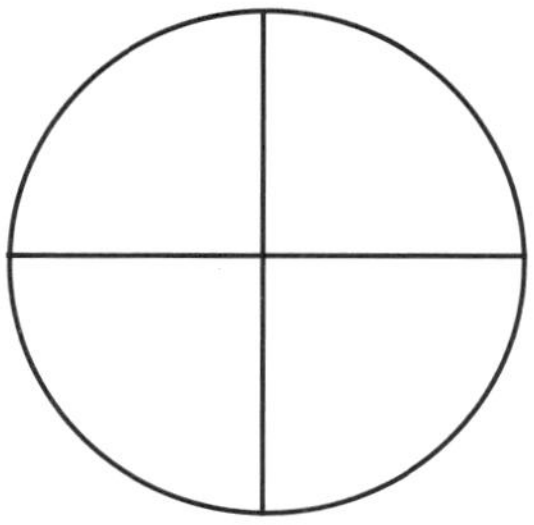

_______ equal parts

Each part is $\frac{\square}{4}$.

3.

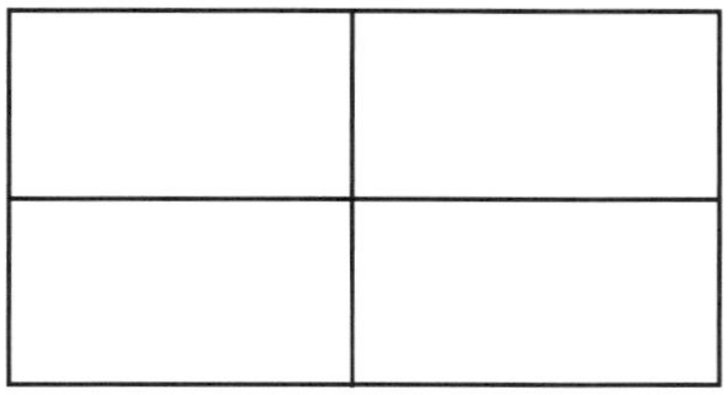

_______ equal parts

Each part is $\frac{1}{\square}$.

4.

_______ equal parts

Each part is $\frac{1}{\square}$.

5.

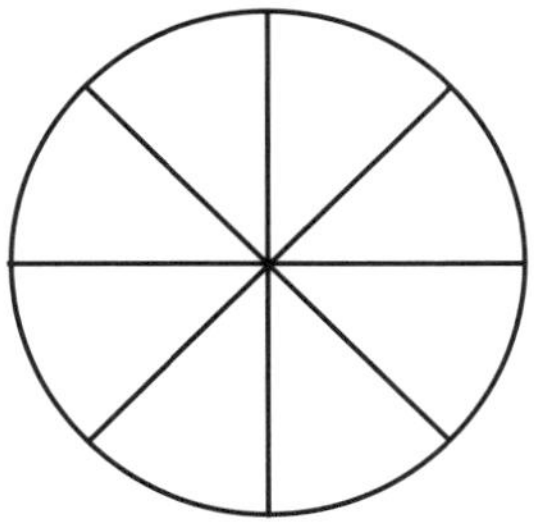

_______ equal parts

Each part is $\frac{1}{\square}$.

6.

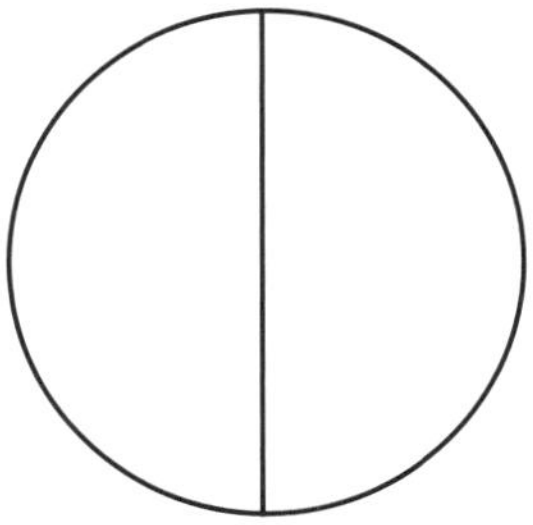

_______ equal parts

Each part is $\frac{1}{\square}$.

☆ **Tell how you found the missing numbers.**

Name ______________________ **Date** ________

Write the missing numbers.

1. 4 equal parts

   $\frac{\square}{4}$ is shaded.

2. 4 equal parts

   $\frac{\square}{4}$ is shaded.

3. ______ equal parts

   $\frac{\square}{\square}$ is shaded.

4. ______ equal parts

   $\frac{\square}{\square}$ is shaded.

Use words to describe the number of shaded parts compared to the whole. Then write the fraction.

5. $\frac{\square}{4}$ is shaded.

   ____ parts out of ____

6. $\frac{\square}{8}$ is shaded.

   ____ out of ____

7. $\frac{\square}{\square}$ is shaded.

   ____ out of ____

8. $\frac{\square}{\square}$ is shaded.

   ____ out of ____

Tell how you found the missing numbers.

Name ______________________________ **Date** __________

Solve.

1. We cut the grapefruit in 2 equal pieces. I ate 1 piece. How much did I eat?

2. A pizza has 8 slices. 3 slices have olives. What fraction of the pizza has olives?

Circle the letter for the correct answer.

3. The book has 8 pages. 5 pages have pictures. What fraction of the book's pages have pictures?

   A $\frac{1}{8}$ of the book

   B $\frac{1}{5}$ of the book

   C $\frac{8}{5}$ of the book

   D $\frac{5}{8}$ of the book

4. Julie is riding her bicycle four miles to the library. After one mile, what fraction of the ride has she traveled?

   A $\frac{3}{4}$ of the ride

   B $\frac{1}{2}$ of the ride

   C $\frac{1}{4}$ of the ride

   D $\frac{4}{8}$ of the ride

Unit 6 Mini-Lesson
Fractions of a Group

Standard

Number and Operations

2.1A (PS) Apply mathematics to problems arising in everyday life, society, and the workplace.

2.3A (SS) Partition objects into equal parts and name the parts, including halves, fourths, and eighths, using words.

Model the Skill

Hand out counters and draw a row of 4 hearts on the board or on paper.

- **Say:** *This picture shows a group of hearts. A group is a set. We can write a fraction that describes part of a set.*
- Shade one of the hearts. **Ask:** *How many parts of the set are shaded?* Allow responses. *How many parts does the set have? What is the total number of parts?* Ask students to count the number of hearts. **Say:** *One-fourth is a fraction. One-fourth describes the red part of the set. One-fourth of the set is red.*
- Now draw four more hearts in the set. **Ask:** *What is the total number of parts in the set now? How many red parts do you see?* Record student responses. **Say:** *One-eighth is a fraction. Now one-eighth describes the red part of the set. One-eighth of the set is red.*
- Invite students to take turns drawing circles or triangles to make a set of two, four, or eight shapes. Ask students to color one part of each set and write a fraction to describe the drawing.

Assess the Skill

Use the following problems to pre-/post-assess students' understanding of the skill.

Ask students to write a fraction to describe the shaded parts of each set.

Name ______________________________ **Date** __________

What fraction of each group is shaded?
Use words to describe the number of shaded parts compared to the whole.

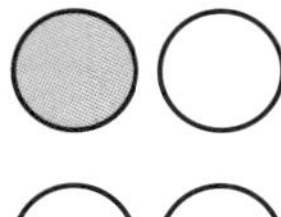

____ parts out of ____

____ out of ____

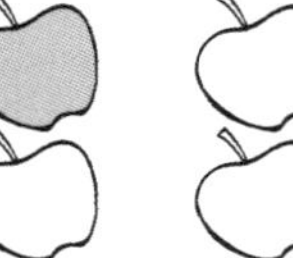

____

____

What fraction of each group is circled?
Write a fraction to describe the circled portion of each set.

5

____

6

____

☆ **Tell how you find the denominator.**

Name ____ **Date** __________

Write a fraction showing one shaded or circled part of each array.

1. 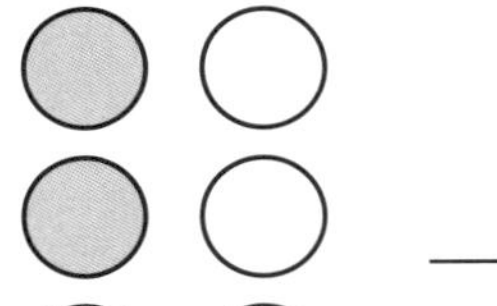 ____

2. 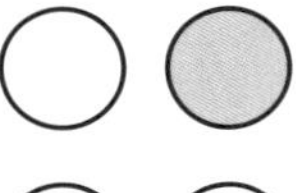 ____

3.  ____

4.  ____

5. 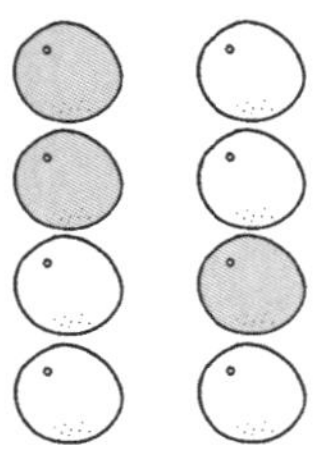 ____

6.  ____

7.   ____

8. 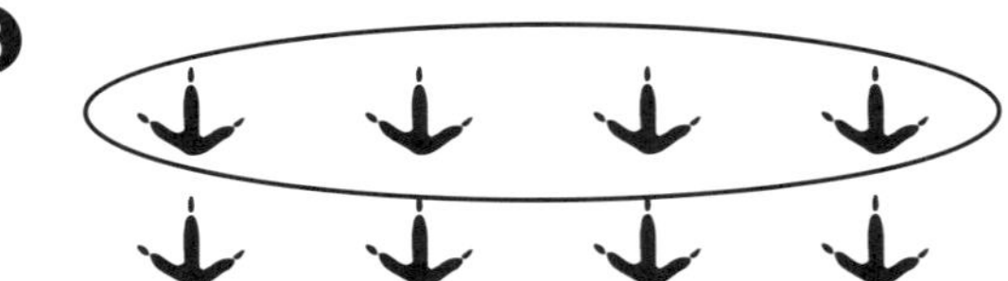 ____

9.     ____

10.  ____

 **Tell how you find the number of equal groups.**

Name ______________________________ Date __________

Write a fraction for each problem.

1. Ella, Jordon, and Everett each have a pet turtle. Harry has a pet fish. What fraction of the group has a pet turtle?

2. The garden has 8 rose bushes. 3 are yellow. The rest are white, red, and pink. What fraction of the rose bushes are yellow?

3. A garage has 8 limousines. 7 of the limousines are black. What fraction of the limousines are black?

4. Cara had 2 boxes of raisins. She ate one box. What fraction of the boxes does she have left?

Choose the correct answer for each problem.

5. There are 4 quarters in a basketball game. After 3 quarters, what fraction of the game is left?

   A $\frac{1}{4}$ of the game

   B $\frac{1}{2}$ of the game

   C $\frac{4}{8}$ of the game

   D $\frac{3}{4}$ of the game

6. There are 8 apples in a bowl. Three apples are green. What fraction of the apples in the bowl is green?

   A $\frac{3}{8}$ of the apples

   B $\frac{1}{4}$ of the apples

   C $\frac{1}{2}$ of the apples

   D $\frac{3}{4}$ of the apples

Unit 7 Mini-Lesson
Use Models to Compare Fractions

Standard

Number and Operations

2.1D (PS) Communicate mathematical ideas, reasoning, and their implications using multiple representations, including symbols, diagrams, graphs, and language as appropriate.

2.3A (SS) Partition objects into equal parts and name the parts, including halves, fourths, and eighths—using words,

Model the Skill

Hand out a rectangular sheet of paper.

- **Say:** *Imagine that you and a friend both want to draw, but you have only one sheet of paper. What can you do so that you each have an equal share of the paper?* (Answers will vary. Possible answer: fold the paper in half and cut along the fold.) Guide students to fold one sheet of paper in half, matching corners. **Say:** *Each share of the paper is the same size. The shares are equal. How is this paper divided?* (into halves) Have students draw a line down the fold to show two equal shares. **Say:** *Each half of the rectangle is one-half. Two halves are the same as one whole.*
- **Say:** *We can fold the paper again to make four equal shares.* Guide students to match corners, fold, and then open the paper to see the four sections. **Say:** *The paper shows fourths. Four-fourths are the same as one whole.* Guide students to draw lines on the new fold to show four equal shares.
- Assign students the appropriate practice page(s) to support their understanding of the skill. If necessary, have students fold another sheet of paper to see the equal shares for halves, fourths, and eighths.

Assess the Skill

Use the following problems to pre-/post-assess students' understanding of the skill.

Ask students to name the number of equal shares in each shape. Then ask them to shade portions of each shape to show the amounts indicated.

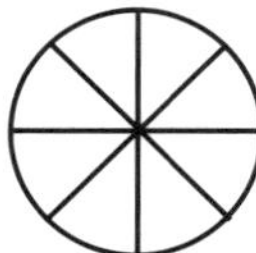

more than 1/2, less than 1

closer to 0 than 1

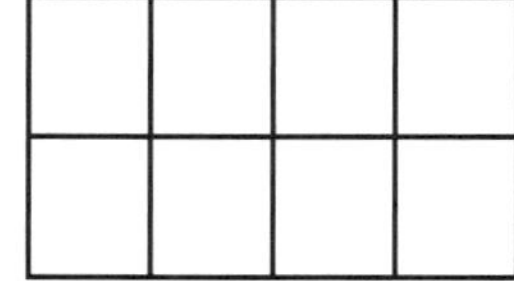

less than 1/2, more than 0

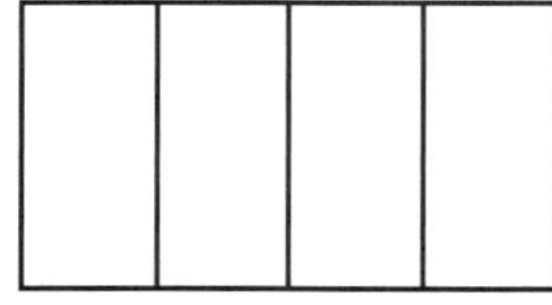

exactly 1/2

Name ______________________________ **Date** __________

1. Show four equal shares.

2. Show two equal shares.

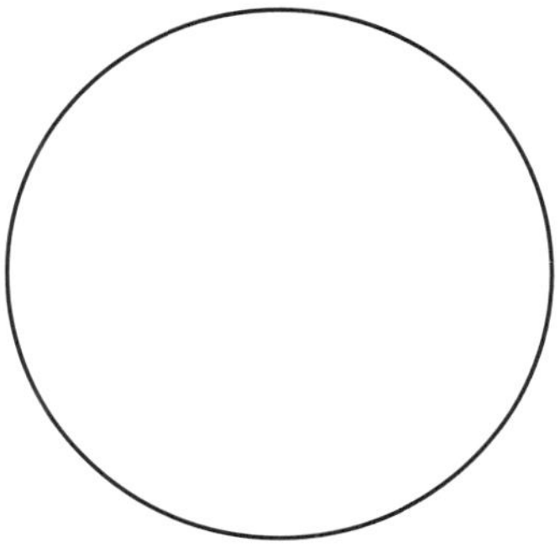

3. Show fourths.

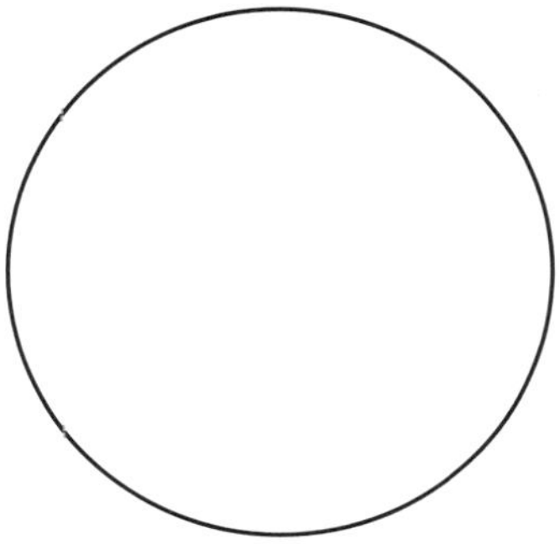

4. Show eighths.

Circle the models that show more than 1/2 shaded.

5.

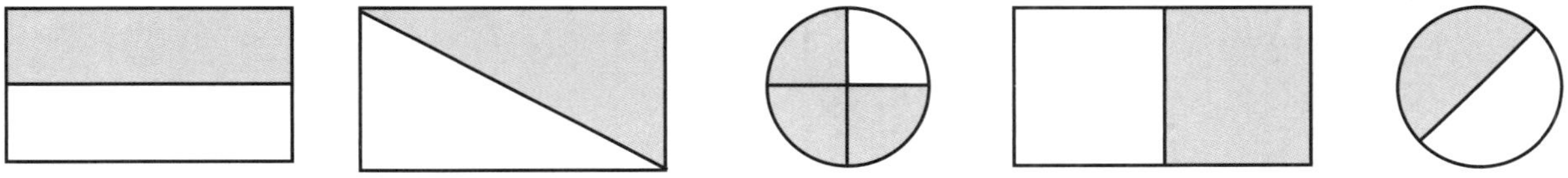

Circle the models that show less than 1/2 shaded.

6.

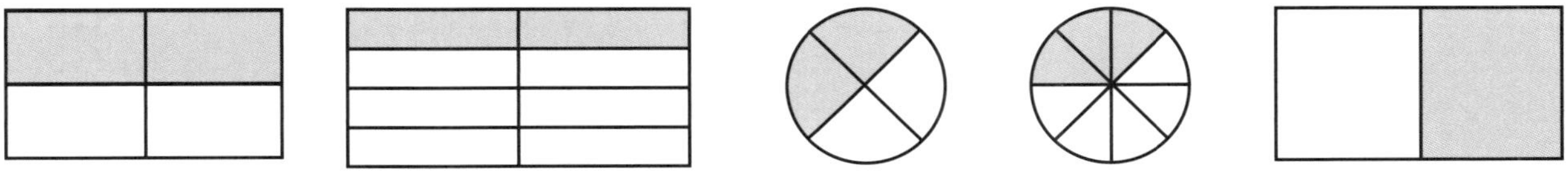

☆ **Tell how equal shares of the same shape can be different shapes.**

Name ______________________________ Date __________

Circle the shapes that show equal amounts. Then label what they show.

1 ____________

2 ____________

3 ____________

4 ____________

5 ____________

6 ____________

7 ____________

8 ____________

Tell how you know which description matches each shape.

Name ______________________________ **Date** __________

Color to show the shares.

1. one-half

2. one-fourth

3. five-eighths

4. two-fourths

5. one-half

6. one-fourth

7. three-eighths

8. two-fourths

9. one-half

10. three-fourths

11. one-eighth

12. three-fourths

☆ **Tell how two-fourths is like one-half.**

Unit 8 Mini-Lesson
Addition and Subtraction Fact Families

Standard

Number and Operations

2.4A (SS) Recall basic facts to add and subtract within 20 with automaticity.

Model the Skill

Hand out connecting cubes and model the following problem on the board.

- **Ask:** *How many of the cubes you see are dark?* (4) *How many lighter cubes are there?* (2) Have students make a cube train with 4 cubes of one color and 2 cubes of another color. **Say:** *The cubes show the addition fact 4 + 2 = 6. Break the cube train into two colors. A related subtraction fact uses the same numbers as the addition fact. What related subtraction fact did you show?* (6 – 2 = 4 or 6 – 4 = 2) Have students record one of the facts.
- **Say:** *Make a cube train for 5 + 3 = ?. What is the total number of cubes?* (8) *Where do you write the total in a subtraction fact?* (Possible answer: It is written first.) Observe as students break their cube train into two colors and record one of the related subtraction facts. (8 – 3 = 5 or 8 – 5 = 3)
- Assign students the appropriate practice page(s) to support their understanding of the skill.

Assess the Skill

Use the following problems to pre-/post-assess students' understanding of the skill.

Have students complete the fact families below.

3 + 1 = 4
____ + ____ = ____
____ – ____ = ____
____ – ____ = ____

6 + 5 = 11
____ + ____ = ____
____ – ____ = ____
____ – ____ = ____

9 + 8 = 17
____ + ____ = ____
____ – ____ = ____
____ – ____ = ____

Name ______________________________ Date __________

Write the related addition and subtraction facts.

1

8 + 2 = 10
____ + ____ = ____
____ − ____ = ____
____ − ____ = ____

2

8 + 4 = 12
____ + ____ = ____
____ − ____ = ____
____ − ____ = ____

3

7 + 8 = 15
____ + ____ = ____
____ − ____ = ____
____ − ____ = ____

4

6 + 13 = 19
____ + ____ = ____
____ − ____ = ____
____ − ____ = ____

5

3 + 9 = 12
____ + ____ = ____
____ − ____ = ____
____ − ____ = ____

6

3 + 7 = 10
____ + ____ = ____
____ − ____ = ____
____ − ____ = ____

7

5 + 10 = 15
____ + ____ = ____
____ − ____ = ____
____ − ____ = ____

8

8 + 12 = 20
____ + ____ = ____
____ − ____ = ____
____ − ____ = ____

9

8 + 6 = 14
____ + ____ = ____
____ − ____ = ____
____ − ____ = ____

 **Tell how you find related addition and subtraction facts.**

Name ______________________________ **Date** __________

Use the numbers shown to write the facts in the fact family.

1. 12 / 5 7

___ + ___ = ___
___ + ___ = ___
___ − ___ = ___
___ − ___ = ___

2. 15 / 7 8

___ + ___ = ___
___ + ___ = ___
___ − ___ = ___
___ − ___ = ___

3. 19 / 12 7

___ + ___ = ___
___ + ___ = ___
___ − ___ = ___
___ − ___ = ___

4. 14 / 5 9

___ + ___ = ___
___ + ___ = ___
___ − ___ = ___
___ − ___ = ___

5.

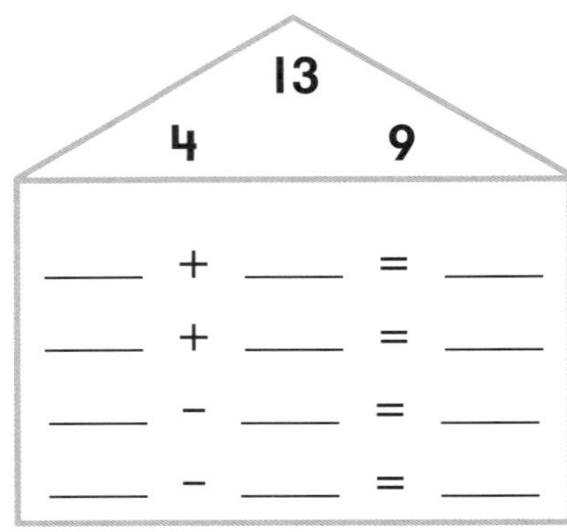

6.

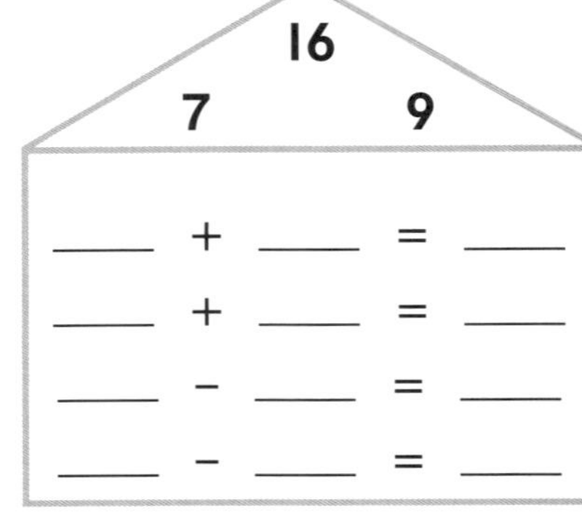

7. 12 / 3 9

___ + ___ = ___
___ + ___ = ___
___ − ___ = ___
___ − ___ = ___

8.

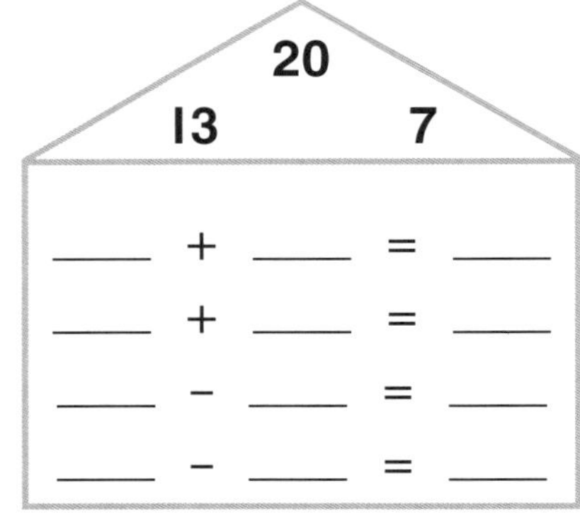

9.

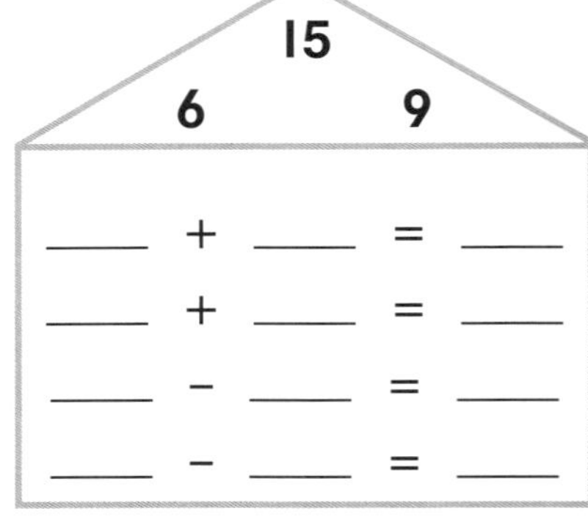

10. 17 / 8 9

___ + ___ = ___
___ + ___ = ___
___ − ___ = ___
___ − ___ = ___

11. 11 / 5 6

___ + ___ = ___
___ + ___ = ___
___ − ___ = ___
___ − ___ = ___

12. 20 / 6 14

___ + ___ = ___
___ + ___ = ___
___ − ___ = ___
___ − ___ = ___

☆ **Explain how you know the facts in a fact family.**

Name ______________________________ **Date** __________

Solve each problem. Show your work.

1. Write three related facts for the number sentence below.

$12 - 5 = 7$

_____ + _____ = _____

_____ + _____ = _____

_____ - _____ = _____

2. Write three related facts for the number sentence below.

$6 + 8 = 14$

_____ + _____ = _____

_____ - _____ = _____

_____ - _____ = _____

Choose the correct answer for each problem.

3. Which sentence is not part of the fact family for 4, 16, and 20?

A $20 + 4 = 24$

B $4 + 16 = 20$

C $20 - 4 = 16$

D $16 + 4 = 20$

$5 + 9 = 14$
$9 + 5 = 14$
$14 - 5 = 9$
___ − ___ = ___

4. Which number sentence completes the fact family above?

A $9 + 5 = 14$

B $14 - 9 = 5$

C $15 - 4 = 9$

D $19 - 5 = 14$

Unit 9 Mini-Lesson

Use Strategies to Add

Standard

Number and Operations

2.4B (SS) Add up to four two-digit numbers and subtract two-digit numbers using mental strategies and algorithms based on knowledge of place value and properties of operations.

2.4C (RS) Solve one-step and multi-step word problems involving addition and subtraction within 1,000 using a variety of strategies based on place value, including algorithms.

Model the Skill

Write the following number line and addition sentence on the board.

25 + 2 =

- **Say:** *You can use different strategies to add. You can count on by 1, 2, or 3. A number line can help you.* Have students use the number line to count on and find the sum. **Say:** *You can draw two jumps to count 2 past 25.* Show students how to draw a curved line from 25 to 26 and then from 26 to 27 to show two jumps. **Ask:** *On what number did you land?* (27) Then write another problem on the board.

  27 + 3 =

- Have students look at the problem. **Ask:** *What number will you circle on the number line?* (27) *How many jumps will you make?* (3) *What is 27 plus 3?* (30)

- Assign students the appropriate activity page(s) to support their understanding of the skill. Remind students using number lines to circle the first addend and draw jumps to the right on the number line equal to the second addend.

Assess the Skill

Use the following problems to pre-/post-assess students' understanding of the skill.

Solve.

16 + 7 =   37 + 8 =   41 + 17 + 19 =   83 + 62 + 45 + 21 =

Name ______________________________ **Date** __________

Add in any order. Show your thinking. Write each sum.

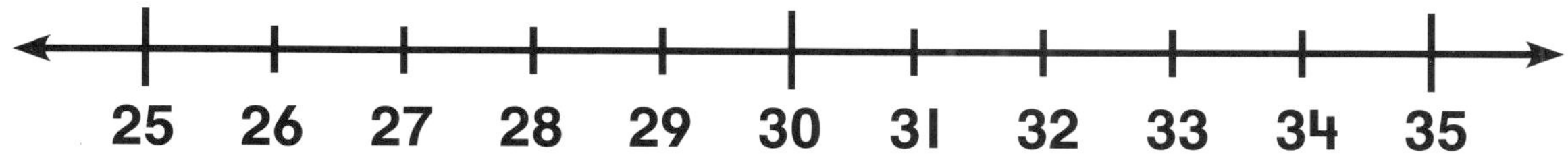

1. 23 + 28 = ______

2. 26 + 14 = ______

3. 56 + 84 = ______

4. 42 + 53 = ______

Circle the numbers you will add first. Show your work. Write the sum.

5. 32 + 15 = ______

6. 39 + 41 = ______

7. 33 + 15 + 55 = ______

8. 87 + 42 + 11 = ______

☆ **Tell how you solved Problem 8.**

Name ______________________________ **Date** __________

Find each sum mentally or with a number line.

1. 41 + 9 = _____

2. 43 + 30 = _____

3. 8 + 23 = _____

4. 16 + 65 = _____

5. 92 + 34 = _____

6. 29 + 28 = _____

7. 76 + 25 + 33 = _____

8. 53 + 37 + 84 = _____

9. 81 + 23 + 14 =

10. 76 + 84 + 95 + 75 = _____

11. 63 + 27 + 81 + 47 = _____

12. 51 + 23 + 79 + 67 = _____

☆ **Tell how you added the four numbers in Problem 11.**

Name ______________________________ **Date** __________

Solve.

1. Jake has 32 cards and Mary has 18 cards. How many cards do they have in all?

2. Catherine has 16 red socks, 18 white socks, and 14 purple socks. How many socks does she have in all?

---

3. Maya has 18 pennies. Her brother gives her 22 more pennies. Then her Dad gives her 61 pennies. How many pennies does she have in all?

4. Rob made 37 cupcakes for the party. Jan brought 28 cupcakes. Eddie brought 42 more. How many cupcakes did they have at the party?

Choose the correct answer for each problem.

5. Jill read 13 pages of her book on Monday. She read 25 more pages on Tuesday. On Wednesday she read another 15 pages. Then she read 17 pages on Friday. How many pages did she read in all?

   A $13 + 25 + 15 + 17 = 60$

   B $13 + 25 + 15 + 17 = 52$

   C $13 + 25 + 15 + 17 = 70$

   D $13 + 25 + 15 + 17 = 78$

6. Ariba and Spencer both have 18 grapes. Then Ariba's mom gives them each 13 more. How many grapes do they have in all?

   A $18 + 18 + 13 = 49$

   B $18 + 13 + 13 = 44$

   C $18 + 18 + 13 + 13 = 60$

   D $18 + 18 + 13 + 13 = 62$

Unit 10 Mini-Lesson
Add Two-Digit Numbers

Standard

Number and Operations

2.4B (SS) Add up to four two-digit numbers and subtract two-digit numbers using mental strategies and algorithms based on knowledge of place value and properties of operations.

2.4C (RS) Solve one-step and multi-step word problems involving addition and subtraction within 1,000 using a variety of strategies based on place value, including algorithms.

Model the Skill

Hand out base-ten blocks and write the following problem on the board.

41 + 26 =

- **Say:** *Let's add these amounts by first joining the tens. How many tens are there in all?* (6) *What is 40 + 20?* Have students write the sum. (60) **Say:** *Let's add the ones. How many ones are there in all?* (7) *What is 1 + 6?* Have students write the sum. (7) **Say:** *Now add the sum of the tens and the sum of the ones. What is the total sum?* (67) Then write another problem on the board.

63 + 15 =

- **Ask:** *How would you show how to add the tens?* (60 + 10) *How would you show how to add the ones?* (3 + 5) Have students show the numbers with blocks, add the tens (70), and add the ones (8). **Say:** *Once you have found the sum of the tens and ones, what do you do?* (Add the tens and ones together.) Write the partial sums and calculate the total sum. (78)
- Assign students the appropriate practice page(s) to support their understanding of the skill. Encourage students to model the addends, adding the tens, adding the ones, and writing the partial sums if necessary.

Assess the Skill

Use the following problems to pre-/post-assess students' understanding of the skill.

Solve.

22 + 47 =

81 + 16 =

63 + 18 =

34 + 58 =

48 + 14 + 32 + 25 =

53 + 14 =

34 + 55 =

58 + 14 =

23 + 47 + 11 =

67 + 12 =

23 + 47 =

72 + 19 =

63 + 12 + 20 =

Name ______________________________ Date __________

Find each sum mentally or with base-ten blocks.

1 54 + 18

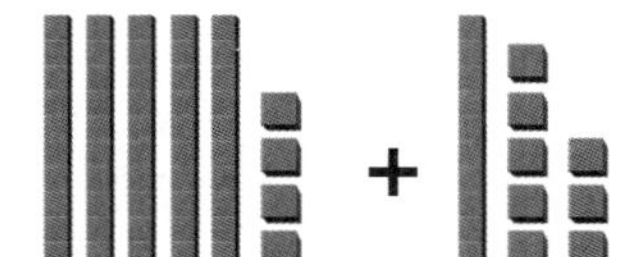

tens	ones
5	4
+ 1	8

2 29 + 37

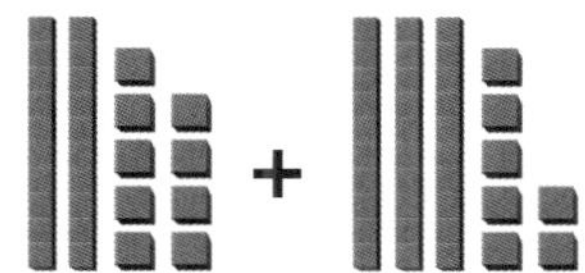

tens	ones
2	9
+ 3	7

3 25 + 73

4 54 + 26

5 46 + 35

6 22 + 69

7 53 + 18 + 32

8 48 + 17 + 25 + 5

Tell how you know when you have to regroup.

Name ______________________________ **Date** __________

Find the sum.

1. 12 + 48 =

2. 55 + 16 =

3. 34 + 22 + 18 =

4. 41 + 13 + 27 =

5. 54 + 16 + 24 =

6. 42 + 13 + 45 =

7. 21 + 57 + 19 =

8. 26 + 43 + 11 =

9. 43 + 27 + 18 + 32 =

10. 37 + 18 + 25 + 11 =

Tell how you found the sum.

Name ______________________________ **Date** __________

Solve.

1. Last season Nina scored 17 goals. This season she scored 14 goals. How many goals has she scored in all?

2. Alex picked 41 apples. Sam picked 29 apples. How many apples did they pick in all?

Choose the correct answer for each problem.

3. Ryan has 14 marbles. Jaya has 25 marbles. Graham has 36 marbles. How many marbles do they have in all?

   A 74

   B 85

   C 65

   D 75

4. Ms. Eves has 22 students in her class. Mr. Jacobs has 25 students. Ms. Repo has 29 students. How many students are in all three classes?

   A 66

   B 76

   C 78

   D 86

Unit 11 Mini-Lesson
Use Strategies to Subtract

Standard

Number and Operations

2.4B (SS) Add up to four two-digit numbers and subtract two-digit numbers using mental strategies and algorithms based on knowledge of place value and properties of operations.

2.4C (RS) Solve one-step and multi-step word problems involving addition and subtraction within 1,000 using a variety of strategies based on place value, including algorithms.

Model the Skill

Draw a number line on the board.

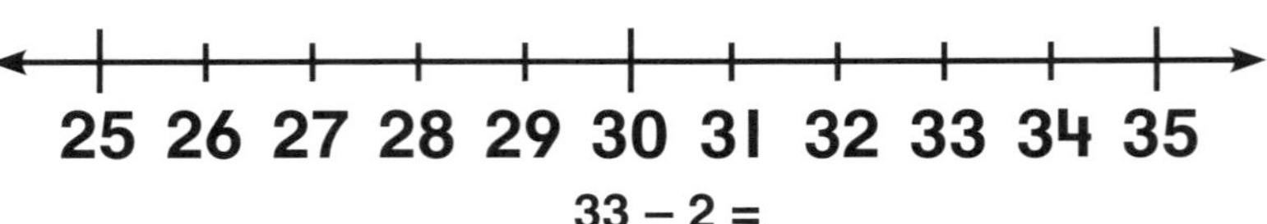

33 – 2 =

- **Say:** *You can use different strategies to subtract. You can count back by 1, 2, or 3 on a number line.* Have students look at the problem and circle the first number in the equation on the number line. **Say:** *You can draw two jumps to count back 2 from 33.* Show students how to draw a curved line from 33 to 32 and then to 31. Encourage students to count back aloud as they draw the jumps. **Ask:** *On what number did you land?* (31) Then write another problem on the board.

32 – 5 =

- Have students repeat with 32 – 5. Observe as students draw the jumps and count back aloud. **Ask:** *What are some other strategies we use when we subtract? Do we think about fact families? Do we use our understanding of place value?* Encourage students to explain their different methods for solving subtraction problems.
- Assign the appropriate practice page(s) to support each student's understanding of the skill.

Assess the Skill

Use the following problems to pre-/post-assess students' understanding of the skill.

13 – 7 =

23 – 7 =

36 – 4 =

47 – 9 =

Name ______________________________ **Date** __________

Use a related addition fact to subtract.

1 17 − 9 = ______

Think:
9 + ? = 17

2 15 − 7 = ______

Think:
7 + ? = 15

3 44 − 8 = ______

Think:
8 + ? = 44

4 31 − 14 = ______

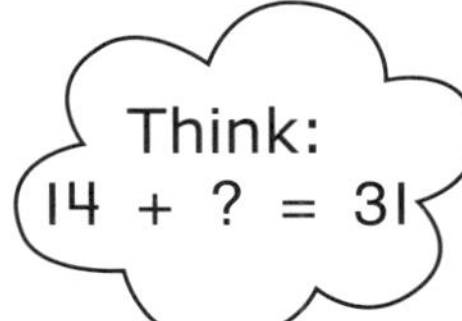

5 58 − 29 = ______

6 66 − 28 = ______

7 32 − 27 = ______

8 73 − 56 = ______

 **Tell how you can use an addition fact to help you subtract.**

Name ______________________________ **Date** __________

Find each difference.

❶ 49 − 4 =

tens	ones

−

❷ 29 − 17

tens	ones

−

❸ 38 − 26 =

tens	ones

−

❹ 75 − 4 =

❺ 84 − 7 − 2 =

❻ 67 − 6 − 11 =

❼ 60 − 34 =

❽ 42 − 28 − 1 =

❾ 37 − 15 − 12 =

❿ 98 − 20 − 10 − 10 =

⓫ 81 − 30 − 5 − 3 =

⓬ 53 − 12 − 11 − 15 =

☆ **Tell how you solved Problem 12.**

Name ______________________________ **Date** __________

Solve.

1. Jude has 84 crayons. He gives 3 to Lily. How many crayons does Jude have left?

2. Maddie has 75 markers. She gives 7 to Carl and 12 to Beth. How many markers does Maddie have left?

Circle the correct answer for each problem.

3. Shen has 91 stamps. He uses 15 stamps to mail some letters. How many stamps does he have left?

   A 76 stamps

   B 86 stamps

   C 77 stamps

   D 31 stamps

4. Farah has 63 cents. She spends 18 cents on Monday and 22 cents on Tuesday. How many cents does she have left?

   A 71 cents

   B 66 cents

   C 55 cents

   D 23 cents

Unit 12 Mini-Lesson

Subtract Two-Digit Numbers

Standard

Number and Operations

2.4B (SS) Add up to four two-digit numbers and subtract two-digit numbers using mental strategies and algorithms based on knowledge of place value and properties of operations.

2.4C (RS) Solve one-step and multi-step word problems involving addition and subtraction within 1,000 using a variety of strategies based on place value, including algorithms.

Model the Skill

Hand out base-ten blocks and write the following problem on the board.

28 – 16 =

- **Say:** *You can use base-ten blocks to model subtraction.* Have students look at the problem and note the subtraction sentence. **Say:** *The beginning number is 28. How do you show 28 with your blocks?* (2 tens, 8 ones) *The number sentence says that we need to subtract 16. How do you show that with the blocks?* (Take away 1 ten and 6 ones.) Guide students to model the subtraction. **Ask:** *What is left?* (12) Observe as students write the difference. Then write another problem on the board.

  30 – 17 =

- **Ask:** *What are you going to show with your blocks to model the subtraction in this problem?* (Possible answer: Show 3 tens). *Can you take away 1 ten and 7 ones from 3 tens?* (no) Show students how to trade 1 ten for 10 ones to show 30 in a different way. **Say:** *Now you have some ones to take away. What is left?* (1 ten, 3 ones; 13)

- Assign the appropriate practice page(s) to support each student's understanding of the skill.

Assess the Skill

Use the following problems to pre-/post-assess students' understanding of the skill.

18 – 17 =	76 – 27 =
34 – 12 =	36 – 4 =
61 – 31 =	47 – 9 =
58 – 43 =	

Name ______________________________ **Date** __________

Find each difference.

 45 − 34 =

	tens	ones
	4	5
−	3	4

2 66 − 43 =

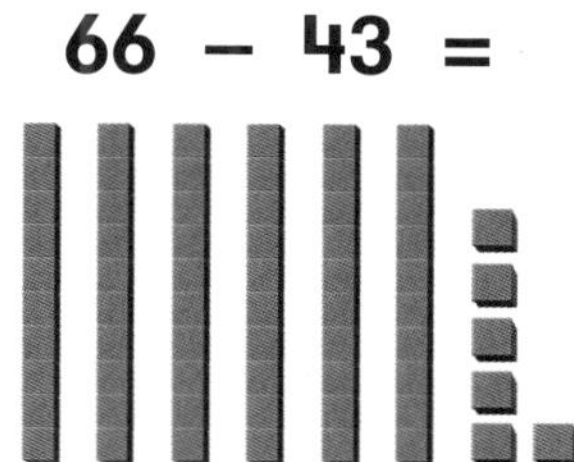

	tens	ones
	6	6
−	4	3

3 59 − 24 =

	tens	ones
−		

4 38 − 18 =

	tens	ones
−		

5 64 − 15 =

	tens	ones
−		

6 74 − 71 =

	tens	ones
−		

7 66 − 20 =

	tens	ones
−		

8 88 − 68 =

	tens	ones
−		

9 30 − 21 − 4 =

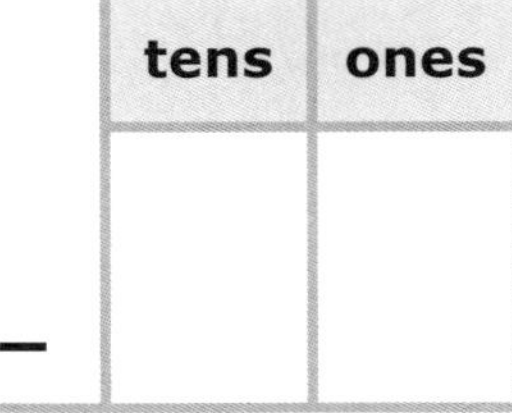

	tens	ones
−		

10 55 − 17 − 18 − 10 =

	tens	ones
−		

 **Tell how you subtract.**

Name ______________________________ **Date** __________

Find each difference.

1. 68 − 47 =

2. 45 − 23 =

3. 39 − 29 =

4. 56 − 23 =

5. 72 − 41 =

6. 67 − 28 =

7. 55 − 29 − 5 =

8. 52 − 23 − 13 =

9. 70 − 48 − 11 =

10. 94 − 37 − 5 − 2 =

11. 65 − 29 − 16 − 11 =

12. 92 − 53 − 22 − 13 =

☆ **Tell how you know when to regroup.**

Name ______________________________ **Date** __________

Solve.

1. Mia has 36 hairpins. She uses 14 in her bun. How many pins does she have left?

2. Rachel has 47 cupcakes. She brings 38 to school. How many cupcakes does she have left?

Circle the correct answer for each problem.

3. The Blue Jays had 48 runs last season. The Red Robins had 29. How many more runs did the Blue Jays have?

   A 9 runs

   B 11 runs

   C 19 runs

   D 21 runs

4. Kyra has 93 dollars. She buys some shoes for 39 dollars, groceries for 23 dollars, and fuel for 18 dollars. How many dollars does she have left?

   A 12 dollars

   B 13 dollars

   C 16 dollars

   D 31 dollars

Unit 13 Mini-Lesson

Solve Word Problems

Standard

Number and Operations

2.1A (PS) Apply mathematics to problems arising in everyday life, society, and the workplace.

2.4C (RS) Solve one-step and multi-step word problems involving addition and subtraction within 1,000 using a variety of strategies based on place value, including algorithms.

Model the Skill

Hand out 10 counters of one color and 10 counters of another color to student pairs. Then write the following problem on the board.

There are 5 red apples and 3 green apples. How many apples are there in all?

- Invite a student to read aloud the problem. **Say:** *You can show a math story problem with objects or pictures. Today we will use counters. How many counters should we show to represent the red apples?* (5) *How many counters should we show to represent the green apples?* (3) Have students use different colors for different color apples.
- **Ask:** *What is the question asking us to find out?* (how many apples there are altogether) *Do you add the groups or compare the groups to find out?* (add) Observe as students put the groups together to find the total number of apples. **Ask:** *What number sentence shows this problem?* (5 + 3 = 8) Observe as students write a number sentence for the problem.
- Assign students the appropriate practice page(s) to support their understanding of the skill.

Assess the Skill

Use the following problems to pre-/post-assess students' understanding of the skill.

There are 16 green apples and 13 yellow apples. How many apples do we have in all?

We eat 12 apples. How many apples are left?

Name ______________________________________ Date _________

Draw a picture to solve. Write the number sentence.

1. Terry has 64 beads. She uses 18 of them on a necklace. How many beads does she have left?

2. Lucas made 29 cookies. Then he made 27 more. How many cookies did he make in all?

3. Ms. Smith bought 72 bottles of water. She gave 12 bottles of water to the other teachers and she gave 30 bottles of water to her students. How many bottles of water does she have left?

4. Sam grew flowers in his garden. He grew 25 blue flowers, 18 red flowers, 37 red flowers, and 2 yellow flowers. How many flowers did he grow in all?

5. Frances has 16 pencils. She buys 20 more. She gives 12 to Sally. How many pencils does she have left?

6. Joe has 23 green stamps and 14 blue stamps. He uses 12 stamps on Monday and 1 stamp on Tuesday. How many stamps does he have left?

 **Tell how you know what number sentence to write.**

Name ______________________________ Date __________

Write a number sentence for each word problem. Then solve.

1. Dia has 14 skirts. She puts 6 in a suitcase. The rest are in her closet. How many skirts are in her closet?

2. Mr. Ramos has 8 ties with stripes and 5 ties with dots. How many ties does he have in all?

3. Charlie has 19 pairs of socks. His father buys him 6 more pairs of socks. How many pairs of socks does Charlie have now?

4. Alex has 12 caps. Cassie has 8 caps. How many more caps does Alex have?

5. Mr. Thorn bought 16 blueberry muffins and 15 banana muffins. His children ate 6 muffins. How many muffins are left?

6. There were 25 rolls on a plate. James ate 4 rolls. Then Isaac ate 5 rolls. How many rolls are left?

7. Mrs. Bella bought 7 pizzas with pepperoni. She bought 4 plain pizzas. Her students ate 3 of the pizzas and the teachers ate 2 of them. How many pizzas are left?

8. There were 20 red pieces and 14 blue pieces on the board. 11 of them were Sam's, 8 of them were Quinn's, and the rest were Steve's. How many pieces on the board were Steve's?

☆ **Tell how you solved Problem 8.**

Name ______________________________ Date ________

Write a number sentence to solve each problem.

1. Sonia and Jay have 35 peaches altogether. Sonia has 22 peaches. How many peaches does Jay have?

2. Wendy had 51 purple beads and 45 green beads. Then she used 23 beads on a bracelet. How many beads does Wendy have now?

3. We have 66 tarts. We have 23 more tarts in the oven. How many tarts do we have in all?

4. Caden noticed 24 math students and 29 science students. If 19 of them were boys, how many were girls?

Circle the correct answer for each problem.

5. Farmer Fred grew 22 melons, 46 oranges and 19 apples. He sells 84 of them at the market. How many items does he have left?

   A 3

   B 4

   C 5

   D 2

6. Rachel's football team scored 37 points and her basketball team scored 45 points. If Rachel scored 12 points in football and 12 points in basketball, how many points did the rest of her teams score?

   A 38

   B 106

   C 58

   D 84

Unit 14 Mini-Lesson
How Much Money?

Standard

Number and Operations

2.5A (RS) Determine the value of a collection of coins up to one dollar.

2.5B (SS) Use the cent symbol, dollar sign, and the decimal point to name the value of a collection of coins.

Model the Skill

Show examples of several coins including pennies, nickels, dimes, and quarters.

- **Ask:** *What types of coins do I have here?* (quarters, dimes, nickels, and pennies) *What is the value of each coin?* (25 cents, 10 cents, 5 cents, and 1 cent) *What is one dime and three pennies?* (13 cents)
- Assign students the appropriate practice page(s) to support their understanding of the skill.

Assess the Skill

Use the following problems to pre-/post-assess students' understanding of the skill.

Ask students to tell or write the amounts shown.

Name ______________________________ **Date** __________

Write each amount. Use the ¢ symbol.

____

2

____

3

____

4

____

5

____

6

____

7

____

8

____

Tell how you count three different kinds of coins.

Name ______________________________ **Date** ________

Match each amount of money to its matching tag.

1.  $3.01

2.  $1.26

3.  $1.60

4.  $0.65

5.  $1.41

6.  $0.40

7.  $2.06

8.  $0.41

 **Tell how you count on to find the total amount of money.**

Name ______________________________ Date __________

Solve.

1. You have two dimes, two nickels, and two pennies. What amount of money do you have?

2. Molly has two quarters, one dime, and one nickel. How much money does Molly have?

Choose the correct answer for each problem.

3. You have two one-dollar bills, a nickel, and two pennies. What amount of money do you have?

   A \$1.52

   B \$2.52

   C \$1.07

   D \$2.07

4. You have one one-dollar bill, three dimes, and five pennies. What amount of money do you have?

   A \$1.35

   B \$1.45

   C \$1.08

   D \$1.53

Unit 15 Mini-Lesson

Model Multiplication

Standard

Number and Operations

2.6A (SS) Model, create, and describe contextual multiplication situations in which equivalent sets of concrete objects are joined.

Model the Skill

Hand out counters and draw a 3 x 2 array of circles on the board or on paper.

- **Say:** *The arrangement of objects in equal rows and columns is called an* ***array****. This is an array. An array has columns and rows. A column goes up and down while a row goes across.* Have students copy the array onto paper. **Say:** *Place a counter on each circle in the first column (or color the column in). How many counters did you use?* (2) Write the number 3 at the bottom of the column. **Say:** *Now place a counter on each circle in the next column. How many counters did you use?* (3) *Write the addition sentence below the counters. What is 3 plus 3?* (6) Observe as students write the sum of 6.
- Draw a 3 x 3 array. **Say:** *The arrangement of objects in equal rows and columns is called an array. How is the first array different from this array?* (Possible answer: It has one more column.) *Place a counter on each circle in a column, or color in and then count how many counters are in each column.* Check that students are placing counters in columns rather than rows. **Ask:** *What is 3 plus 3 plus 3?* (9) *Is that the total number of counters you have used?* (yes)
- Assign students the appropriate practice page(s) to support their understanding of the skill. **Say:** *You can use addition to find the total number of items in an array.*

Assess the Skill

Use the following problems to pre-/post-assess students' understanding of the skill.
Ask students to use repeated addition to find the sum total for each array.

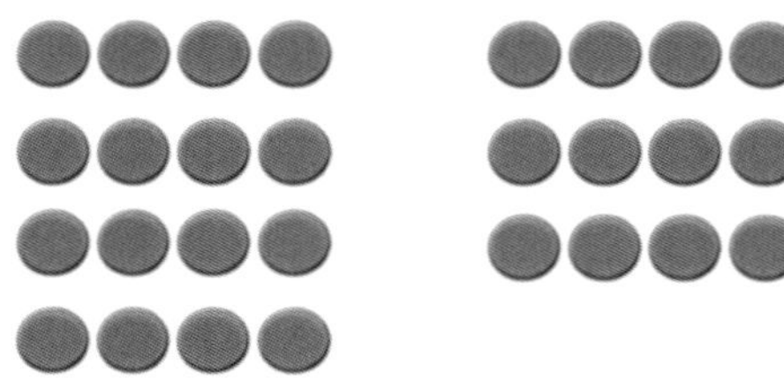

Name ______________________________ **Date** __________

Write the number of items in each column. Add equal groups to find the total.

1 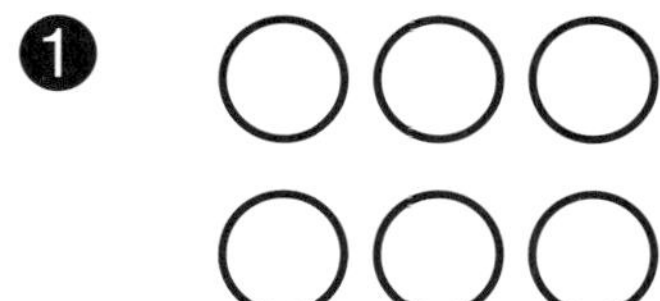

2 + 2 + 2 = _____

2 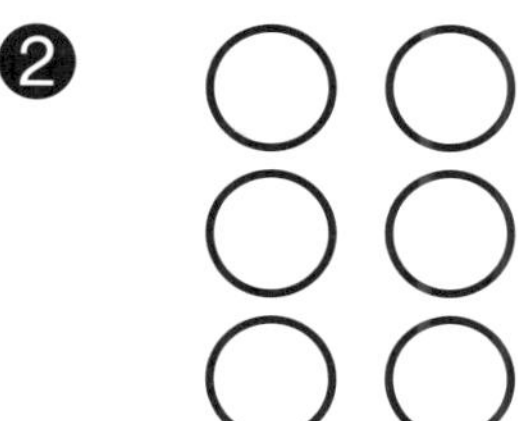

___ + ___ = ___

3 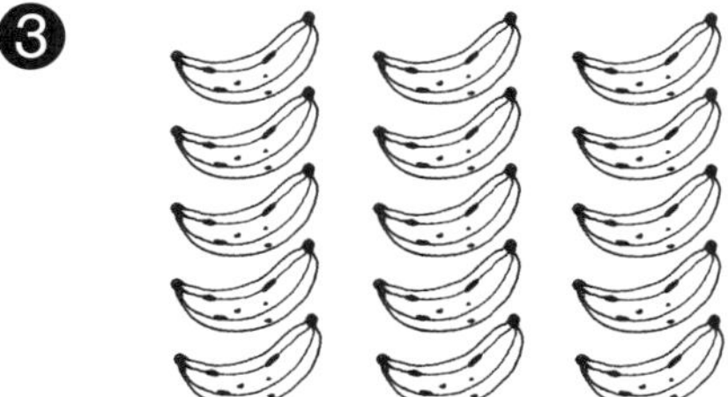

___ + ___ + ___ = ___

4 

___ + ___ + ___ + ___ + ___ = ___

5 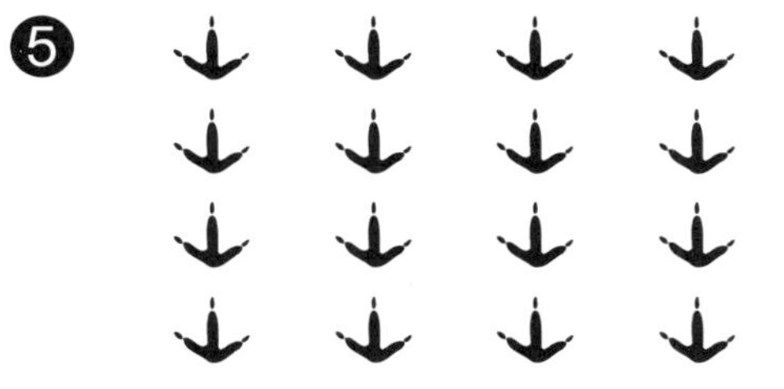

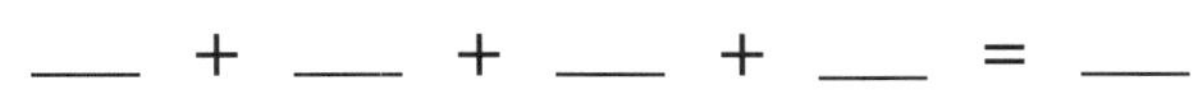

___ + ___ + ___ + ___ = ___

6 

___ + ___ + ___ + ___ = ___

7 

___ + ___ + ___ = ___

8 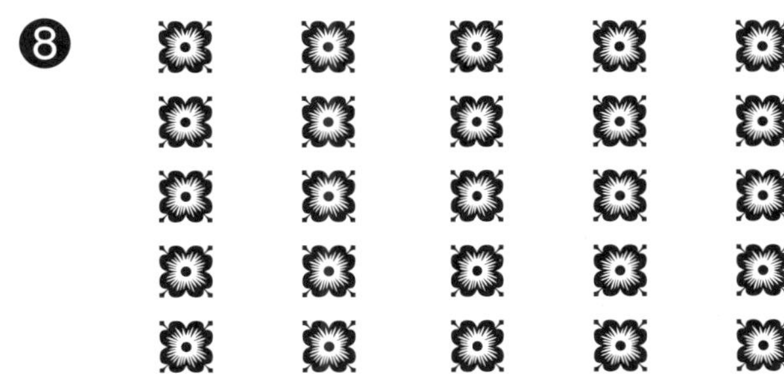

___ + ___ + ___ + ___ + ___ = ___

How do your answers for 6 and 7 compare?
Tell how the arrays are alike and different.

Name ______________________________ **Date** __________

Write a number sentence for each array.

1 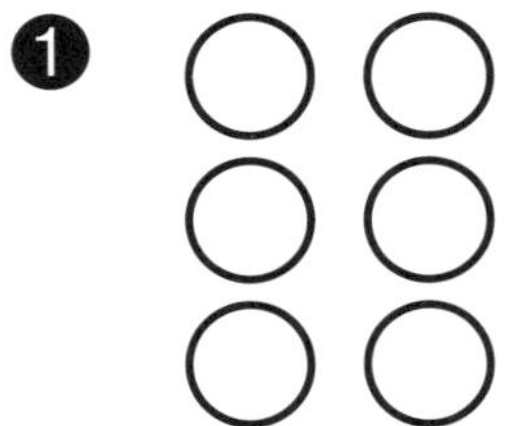

___ + ___ = ___

2 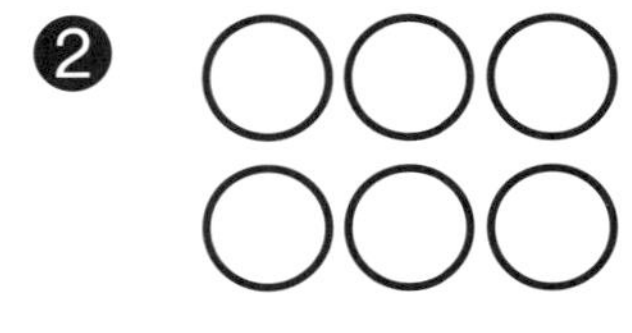

___ + ___ + ___ = ___

3 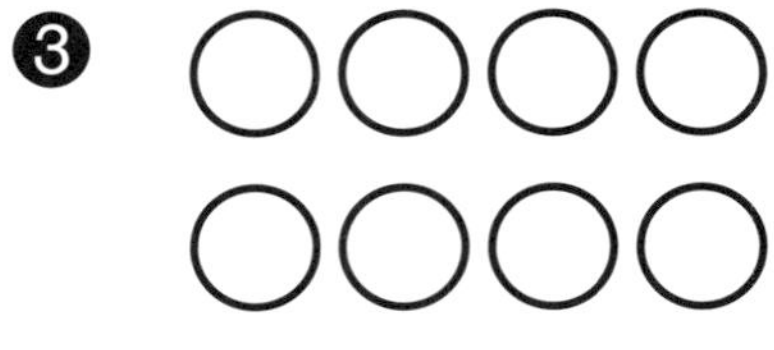

___ + ___ + ___ + ___ = ___

4 

___ + ___ + ___ = ___

5 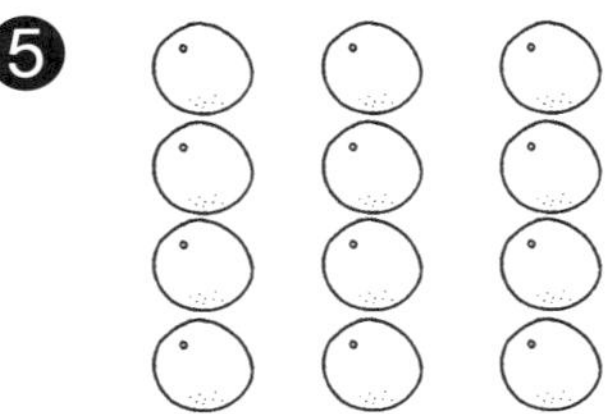

___ + ___ + ___ = ___

6 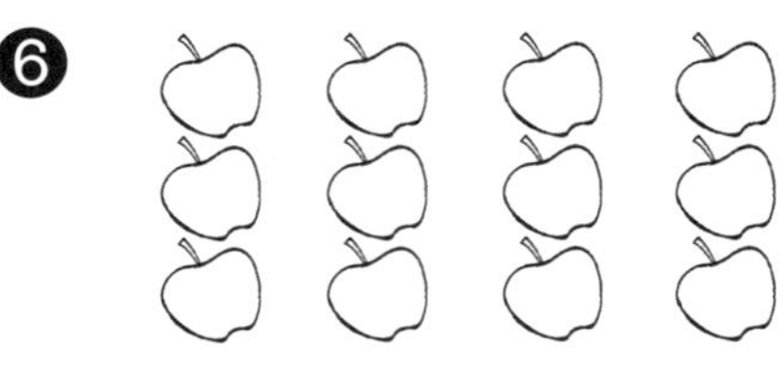

___ + ___ + ___ + ___ = ___

7 

____

8 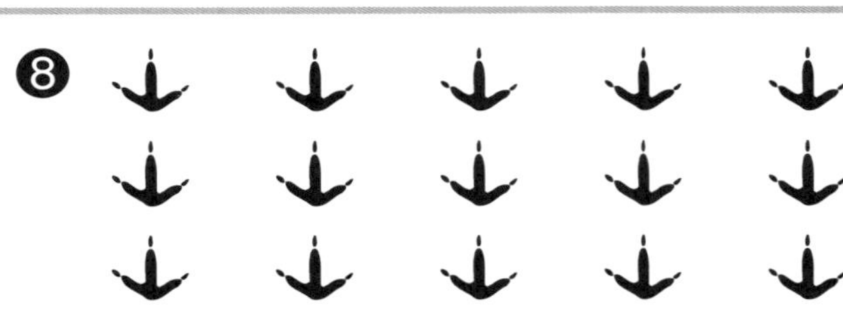

____

9 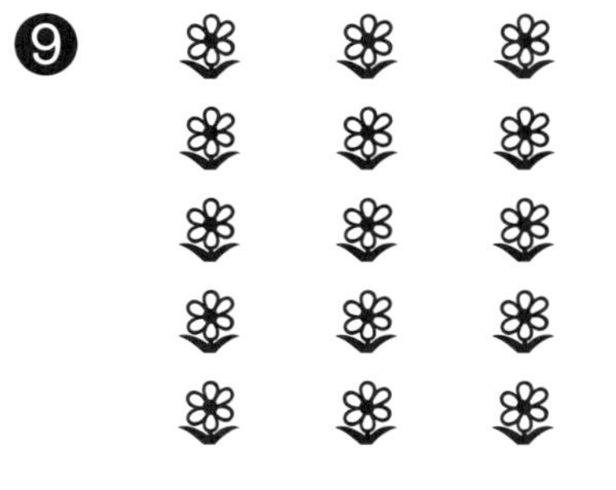

____

10 

____

 **Tell how skip counting can help you find the total number in an array.**

Name ______________________________ Date __________

Draw an array for each problem. Then write a number sentence for each array.

1. Teddy, Meta, and Henry each have 5 dollars. How many dollars do they have in all?

2. The grocer has 3 rows of peppers. Each row has 8 peppers. How many peppers does the grocer have in all?

3. The vegetable garden has 4 rows of tomato plants. Each row has 9 tomato plants. How many tomato plants are there in all?

4. Annabelle has 5 boxes of raisins. Each box has 10 raisins. How many raisins does she have in all?

Choose the correct answer for each problem.

5. There are 3 cartons of eggs. Each carton has 6 eggs. How many eggs are there in all?

   A 9
   B 12
   C 18
   D 36

6. There are 6 bags of apples. There are 7 apples in each bag. How many apples are there in all?

   A 13
   B 26
   C 42
   D 35

Unit 16 Mini-Lesson
Model Division

Standard

Number and Operations

2.6B (SS) Model, create, and describe contextual division situations in which a set of concrete objects is separated into equivalent sets.

Model the Skill

Hand out counters and draw a 3 x 2 array of circles on the board or on paper.

- **Say:** *The arrangement of objects in equal rows and columns is called an array. This is an array. An array has columns and rows. A column goes up and down while a row goes across.* Have students copy the array onto paper. **Say:** *Place a counter on each circle (or color the circles in). How many counters did you use?* (6) **Say:** *How many columns do you see?* (3) *How many rows?* (2) *Each row is an equal group. Each column is an equal group, too. If I divide 6 counters into 3 equal groups, how many groups would I have?* (2). *If I divide 6 counters into 2 equal groups, how many groups would I have?* (3)
- Draw a 3 x 3 array. **Say:** *The arrangement of objects in equal rows and columns is called an array. How is the first array different from this array?* (Possible answer: It has one more column.) *Place a counter on each circle in a column, or color in and then count how many counters are in each column.* Check that students are placing counters in columns rather than rows. **Ask**: *If I divide 9 counters into 3 equal groups, how many groups would I have?* (3)
- **Say:** *You can use repeated subtraction to find the number of equal groups in an array.*

Assess the Skill

Use the following problems to pre-/post-assess students' understanding of the skill.

Ask students to use repeated subtraction to find to find the number of equal groups.

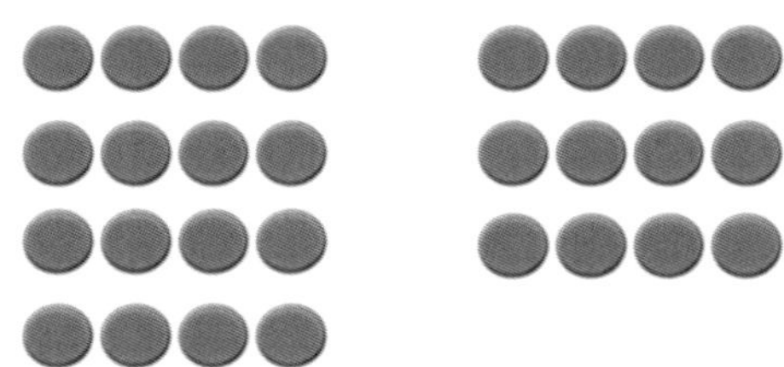

Name ______________________________ **Date** __________

Find how many equal groups are in each amount.

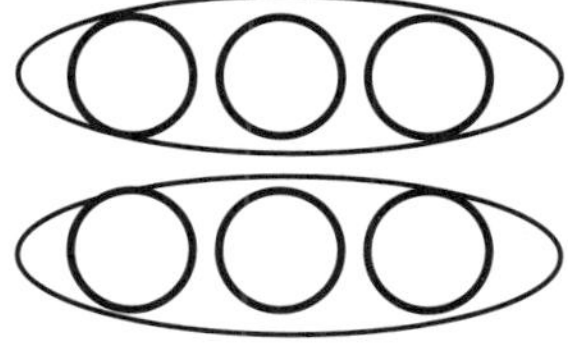

6 − 3 − 3 = 0

______ equal groups of **3**

______ equal groups of **2**

2
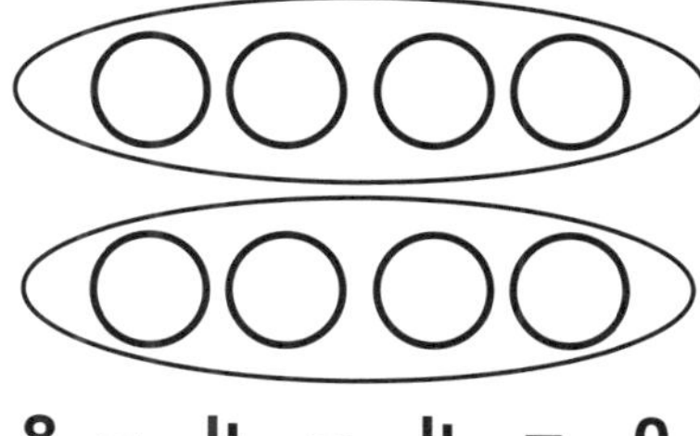

8 − 4 − 4 = 0

______ equal groups of **4**

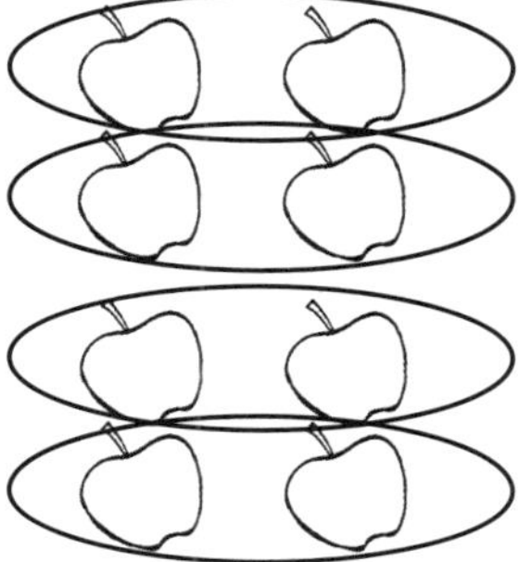

8 − 2 − 2 − 2 − 2 = 0

______ equal groups of ______

4

10 − 5 − 5 = 0

______ equal groups of ______

5
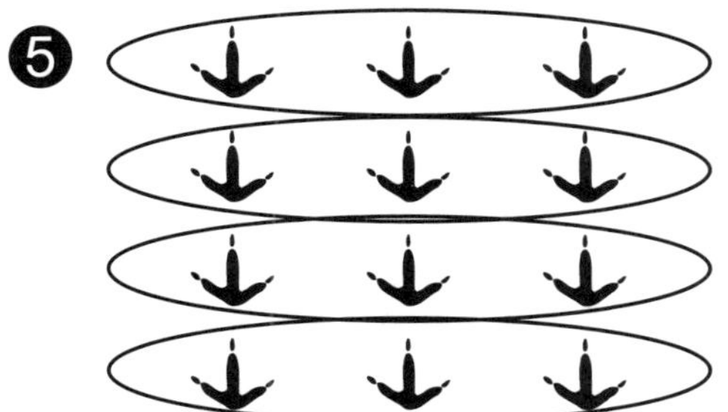

12 − 3 − 3 − 3 − 3 = ____

______ equal groups of ______

6
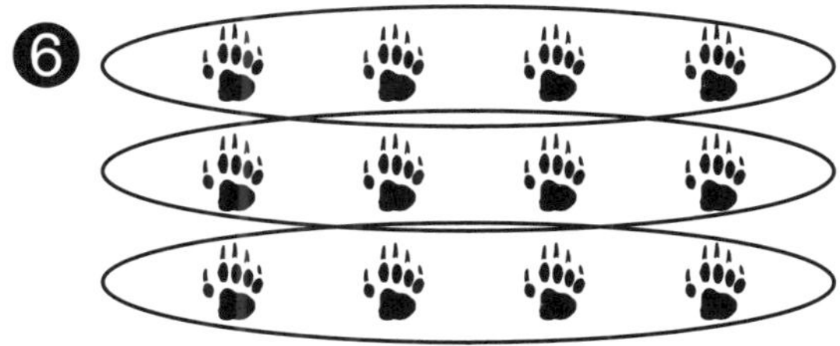

12 − 4 − 4 − 4 = ____

______ equal groups of ______

How do your answers for 5 and 6 compare? Tell how the arrays are alike and different.

Name ______________________________ **Date** __________

Write the number of equal groups and the number in each group.

1

____

____

2

____

____

3

____

____

4

____

____

5

____

____

6

____

____

7

____

____

8

____

____

9

____

____

10

____

____

☆ **Tell how you found the number of equal groups for each array.**

Name ______________________________ Date ________

Write a number sentence and draw an array for each problem.

1. Four friends have 20 dollars among them. If they all have the same amount of money, how many dollars does each friend have?

2. The grocer has 15 cans of peppers. If she places 5 cans in each row, how many rows can she make?

3. The vegetable garden has 9 tomato plants. There are 3 rows of tomato plants. How many tomato plants are in each row?

4. Peter has 16 pretzels. If he gives 2 pretzels to each friend, how many friends can he give pretzels to?

Choose the correct answer for each problem.

5. There are 18 eggs. Each carton has 6 eggs. How many cartons are there in all?

   A 2
   B 3
   C 4
   D 6

6. There are 16 apples in 4 bags. Each bag has the same number of apples. How many apples are in each bag?

   A 2
   B 3
   C 4
   D 6

Unit 17 Mini-Lesson

Find Patterns

Standard

Mathematical Process Standards

2.1F (PS) Analyze mathematical relationships to connect and communicate mathematical ideas.

Model the Skill

Hand out counters and write the following problem on the board.

$$2 + 2 = 4$$
$$2 + 2 + 2 = 6$$

- Have students use counters to model the problems. **Say:** *Today we are going to look for patterns when we add.* Have students model along as you demonstrate how to count by 2s.
- **Ask:** *How many groups of counters are there? How many counters are there in the group?*
- **Say:** *Now let's try adding other equal groups. What pattern do you see when we add by 3s?*
- Follow a similar process with 5s and 10s. Ask students to describe the patterns they see.
- Assign students the appropriate practice page(s) to support their understanding of the skill.

Assess the Skill

Use the following problems to pre-/post-assess students' understanding of the skill.

Write the following patterns on the board and ask students to complete each pattern and then define the rule for each pattern.

2, 4, 6, ____, ____, 12, ____, 16, ____, 20

3, 6, ____, ____, 15, ____, 21, ____, 27, ____, 33, ____

16, 20, ____, 28, ____, ____, 40, 44

12, 18, ____, 30, ____, 42, ____, ____

Name ______________________________ **Date** __________

Complete the hundreds chart.

1	**2**	**3**		**5**	**6**	**7**	**8**		**10**
	12		**14**		**16**		**18**		
21		**23**		**25**		**27**		**29**	
	32							**39**	
		43	**44**						**50**
51				**55**					
					66		**68**		
71						**77**			
				85			**88**		
91								**99**	**100**

 **Write about the patterns you see in the table.**

Name ____________________ **Date** ________

Complete the pattern for each totals. Then write the rule.

1.

Input	Output
2	12
5	15
7	17
13	23

rule: ____________

____

2.

Input	Output
3	11
6	14
9	17
12	20

rule: ____________

____

3.

Input	Output
4	16
6	18
8	20
10	

rule: ____________

____

4.

Input	Output
1	12
3	14
5	16
7	18

rule: ____________

____

5.

Input	Output
45	37
40	32
35	27

rule: ____________

____

6.

Input	Output
58	38
46	26
34	14

rule: ____________

____

7.

Input	Output
37	74
40	80
43	86
46	

rule: ____________

____

8.

Input	Output
68	34
58	29
48	24
38	

rule: ____________

____

9.

Input	Output
30	44
27	
24	
21	35

rule: ____________

____

10.

Input	Output
13	31
14	33
15	35
16	

rule: ____________

____

Name ______________________________ **Date** __________

Solve.

1. Which column in the hundreds chart has the numbers 9, 29, 59, and 89?

2. Which row in the hundreds chart has the numbers 16, 12, 18, and 20?

3. Describe the pattern in the table below.

Number of Trucks	1	2	3	4	5	6
Number of Tires	10	20	30			

Circle the letter for the correct answer.

4. There are 6 bookshelves. Each bookshelf has 7 books. Which expression can be used to show the total number of books?

A 7 + 6
B 7 + 7 + 7 + 7 + 7 + 7
C 6 + 6 + 6 + 6 + 6 + 6
D 7 + 6 + 7 + 6

5. There are 8 wheels on each stroller. There are 5 strollers in the park. How many stroller wheels are in the park?

A 20 wheels
B 40 wheels
C 60 wheels
D 80 wheels

Unit 18 Mini-Lesson

Make a Table

Standard

Mathematical Process Standards

2.1F (PS) Analyze mathematical relationships to connect and communicate mathematical ideas.

Model the Skill

Show the following data table.

- **Say:** *You can use a table to show data. We can also look at tables and find patterns within the data. The table here shows the number of bicycles and the number of tires.* **Ask:** *According to the table, how many tires would be on 5 bicycles?* (10) *What is the rule for this pattern?* (add 2 tires for each 1 bicycle)
- **Say:** *The next table shows the number of cars loaded on a transport truck.* **Ask:** *According to the table, how many cars would fit on 4 transport trucks?* (36) *What is the rule for this pattern?* (add nine cars for each 1 truck)
- Assign students the appropriate practice page(s) to support their understanding of the skill.

Number of Bicycles	Number of Tires
1	2
2	4
3	6
4	8
5	?

Number of Trucks	Number of Cars
1	9
2	18
3	27
4	?
5	45

Assess the Skill

Use the following activity to pre-/post-assess students' understanding of the skill.

- Using items such as vehicles equipped with wheels such as bicycles, tricycles, cars, trucks, ask students to make a table and generate patterned data.

Name ________________________________ Date ________

Complete each table.

Number of Bicycles	Number of Tires
1	2
2	
3	
4	
5	
6	
7	
8	
9	
10	

Number of Tricycles	Number of Tires
1	3
2	
3	
4	
5	
6	
7	
8	
9	
10	

Number of Cars	Number of Tires
1	4
2	
3	
4	
5	
6	
7	
8	
9	
10	

Number of Trucks	Number of Tires
1	6
2	
3	
4	
5	
6	
7	
8	
9	
10	

Name ______________________________ **Date** __________

Look for patterns. Then complete each table.

1.

Number of Bicycles	1	2	3	4	5	6
Number of Wheels	2	4	6			

2.

Number of Nickels	1	2	3	4	5	6
Number of Cents	5	10	15			

3.

Number of Chairs	1	2	3	4	5	6
Number of Legs	4	8				

4.

Number of Insects	1	2	3	4	5	6
Number of Legs	6		18			

5.

Number of Gallons	1	2	3	4	5	6
Number of Pints	8	16			40	

6.

Number of Centimeters	1	2	3	4	5	6
Number of Millimeters	10	20	30			

7.

Number of Tricycles	0	1	2	3	4	5
Number of Wheels	0	3				

8.

Number of Skateboards	0	2	4	6	8	10
Number of Wheels	0	8				

Tell about the patterns in Table 8.

Name ______________________________ **Date** __________

Solve.

1. Sam has more than 50 cars in his toy car collection. He has started to make a table that shows the number of cars and the number of wheels. Complete the table.

Number of Cars	Number of Wheels
2	
4	
6	
8	
10	

2. Which rule best describes the pattern in Sam's table?

   A Add 2 wheels

   B Subtract 2 wheels

   C Add 8 wheels

   D Add 4 cars

3. How many wheels would 11 cars have?

   A 40

   B 36

   C 48

   D 44

Unit 19 Mini-Lesson
Extend Patterns

Standard

Mathematical Process Standards

2.1F (PS) Analyze mathematical relationships to connect and communicate mathematical ideas.

Model the Skill

- **Draw the following figures on the board.**

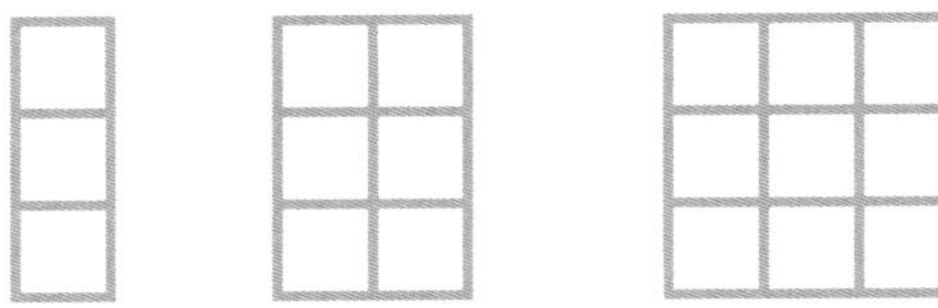

- **Say:** *We are going to look for patterns today. Look at these figures made from squares. How is the second figure different from the first?* (It has 3 more squares.) *How is the third figure different from the second?* (It has 3 more squares.) *How many squares should the next figure have?* (12)
- **Ask:** *How does the rule "add 3" help us understand the pattern?* (It describes how the pattern increases.) Have a volunteer extend the pattern by drawing the next two figures. (3 x 4 squares; 3 x 5 squares)
- Assign students the appropriate practice page(s) to support their understanding of the skill.

Assess the Skill

Use the following problems to pre-/post-assess students' understanding of the skill.

- **Ask:** *How would you complete the following patterns?*

  1, 3, 7, 15, 31, ____, ____, ____

  108, 90, 72, ____, ____, ____

Name ______________________________ Date __________

Follow the rule. Extend the pattern.

 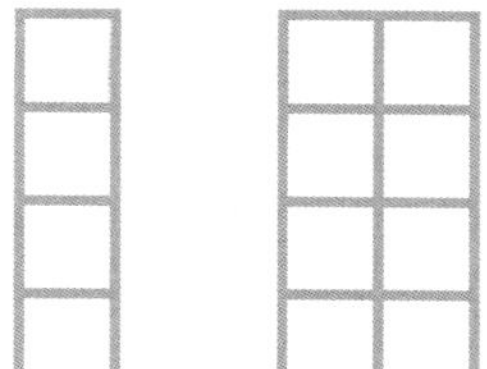 

The pattern is ____________

 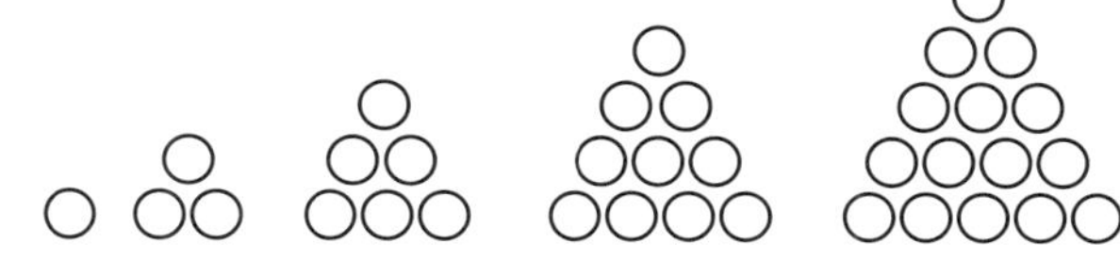

The pattern is ____________

---

3 2, 4, 8, ______, ______, ______

The pattern is ____________

4 1, 3, 5, 7, ______, ______

The pattern is ____________

---

5 24, 36, 48, ______, ______, ______

The pattern is ____________

6 99, 88, 77, 66, ______, ______

The pattern is ____________

---

7 2, 5, 8, 11

The pattern is ____________

8 3, 6, 12, 24

The pattern is ____________

 **Tell how the rule describes the pattern.**

Name ______________________________ **Date** __________

Describe the pattern. Complete the pattern for each problem.

1

2

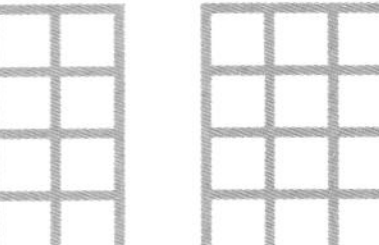

3

7, 14, 28, 56, ______, ______

4

600, 580, 560, ______, ______

5

78, 65, 52, 39, ______, ______

6

105, 200, 295, ______, ______

7

Input	Output
2	10
3	15
4	20

8

Input	Output
2	3
3	4
4	5

9

Input	Output
5	15
10	30
15	
	60

10

Input	Output
2	10
4	20
6	30
8	

 **Create your own pattern and define the rule.**

Name ______________________________ **Date** __________

Solve.

1. Use the following rule to make a pattern.

   Rule: Add 9

   ____

2. Use the following rule to make a pattern.

   Rule: Subtract 4

   ____

3. Complete the pattern, then define the rule for the pattern.

   1, 6, 11, 16, 21, ___, ___

   Rule: ______________________

4. Complete the pattern, then define the rule for the pattern.

   64, 53, 42, 31, ___, ___

   Rule: ______________________

Circle the letter for the correct answer.

5. Which of the following numbers completes the pattern below?

   125, 100, 75, 50, ___, ___

   A 10, 5

   B 25, 5

   C 10, 2

   D 25, 0

6. Which of the following best describes the pattern below?

Input	Output
20	15
18	13
16	11
14	9

   A add 10

   B subtract 2

   C subtract 5

   D add 5

Unit 20 Mini-Lesson
Describe Plane Figures

Standard

Geometry and Measurement

2.8A (SS) Create two-dimensional shapes based on given attributes, including number of sides and vertices.

Model the Skill

Hand out tangrams and cubes.

- Display models of triangles, quadrilaterals, pentagons, hexagons, and cubes. **Ask:** *Which one of these shapes is most different from the other shapes?* (a cube) Hold up a cube and explain that it is a three-dimensional shape while the other shapes are two-dimensional, or flat, shapes. **Ask:** *How many flat sides, or faces, does a cube have?* Have two volunteers work together to determine the number of faces. (6) **Ask:** *What shape are the faces of a cube?* (square)
- Hold up a triangle. **Say:** *Use a blue crayon to trace the sides of the triangle. How many sides did you trace?* (3) Allow tactile learners to touch the triangle models. **Say:** *To count the angles, use a red crayon to make a mark on each angle. How many angles are there?* (3) Explain to students that ***tri-*** means ***three***, and a triangle is a shape with three angles.
- Have students trace and mark the sides of a quadrilateral. **Say:** *Think about other words that you know that begin with **quad**.* Quad *means **four**. A quadrilateral has four sides and four angles.* Point out various quadrilaterals including a square, rectangle, rhombus, trapezoid, and other parallelograms.
- Assign students the appropriate practice page(s) to support their understanding of the skill. Guide students to link the names of each shape to the number of sides and angles of each shape.

Assess the Skill

Use the following problems to pre-/post-assess students' understanding of the skill.

Have students pick up handfuls of tangram shapes and ask them to trace and name each one.

Name ______________________________ Date __________

Match each shape to its name.

❶ parallelogram

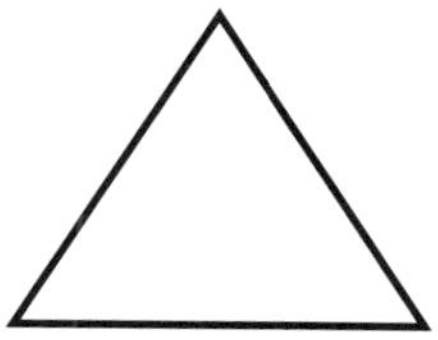

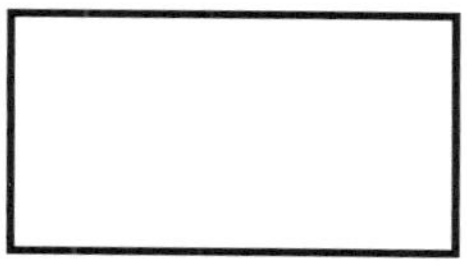

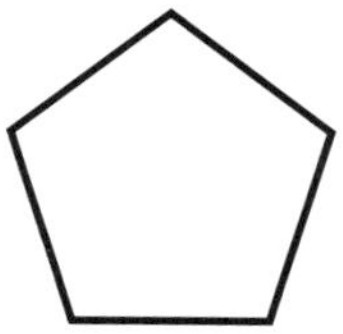

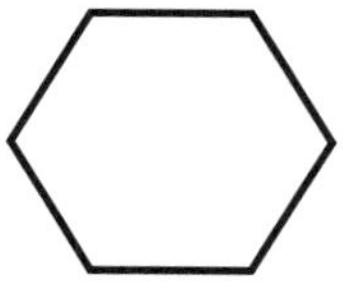

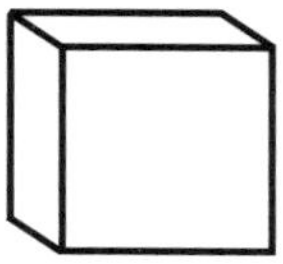

❷ triangle

❸ rectangle

❹ cube

❺ hexagon

❻ pentagon

 **Circle all of the quadrilaterals.**

Name ____________________ Date ________

Label each shape.

1 ____________

2 ____________

3 ____________

4 ____________

5 ____________

6 ____________

7 ____________

8 ____________

9 ____________

10 ____________

11 ____________

12 ____________

☆ **Tell how you know which shapes are quadrilaterals.**

Name ______________________________ Date __________

Draw each shape described. Write its name.

1. I have four equal sides and four equal angles. What am I?

____

2. I have six sides and six angles. What am I?

____

3. I have four sides and four angles. What am I?

____

4. I have six faces and six angles. What am I?

____

5. I have three sides and three angles. What am I?

____

6. I have five sides and five angles. What am I?

____

☆ **Tell how you know which shape is described.**

Unit 21 Mini-Lesson

Describe Solid Figures

Standard

Geometry and Measurement

2.8B (RS) Classify and sort three-dimensional solids, including spheres, cones, cylinders, rectangular prisms (including cubes as special rectangular prisms), and triangular prisms, based on attributes using formal geometric language.

Model the Skill

Hand out tangrams and solid figures.

- Display models of triangles, quadrilaterals, pentagons, hexagons, and cubes. **Ask:** *Which one of these shapes is most different from the other shapes?* (a cube) Hold up a cube and explain that it is a three-dimensional shape while the others are two-dimensional, or flat, shapes. **Ask:** *How many flat sides, or faces, does a cube have?* Have two volunteers work together to determine the number of faces. (6) **Ask:** *What shape are the faces of a cube?* (square)
- Repeat with other three-dimensional figures such as cones, spheres, rectangular prisms, triangular pyramids, etc.
- Assign students the appropriate practice page(s) to support their understanding of the skill. Guide students to link the names of each shape to the number of sides and angles of each shape.

Assess the Skill

Use the following problems to pre-/post-assess students' understanding of the skill.

Have students pick up solid figures and ask them to describe and name each one. Then have them list examples of household or classroom objects that can be classified as each one of the figures.

Name ______________________________________ Date __________

List the number of sides for each shape. Then list the number of angles.

❶ **pyramid**

sides ________
angles ________

❷ **rectangular prism**

sides ________
angles ________

❸ **cylinder**

sides ________
angles ________

❹ **sphere**

sides ________
angles ________

❺ **cube**

sides ________
angles ________

❻ **cone**

sides ________
angles ________

☆ **Tell how the number of sides matches the shape name.**

Name ______________________ **Date** __________

Label each shape.

1 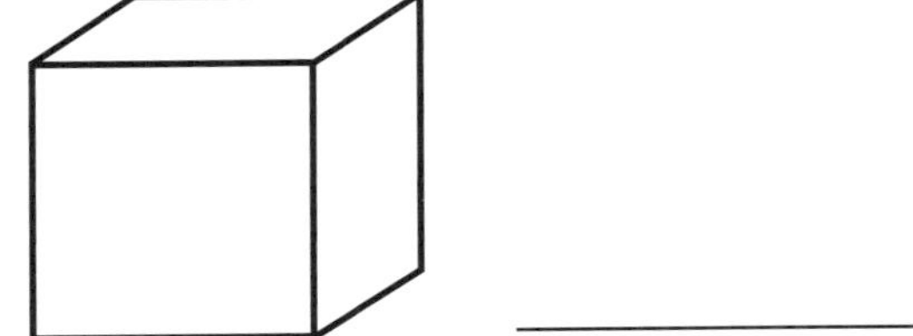 ______________

2 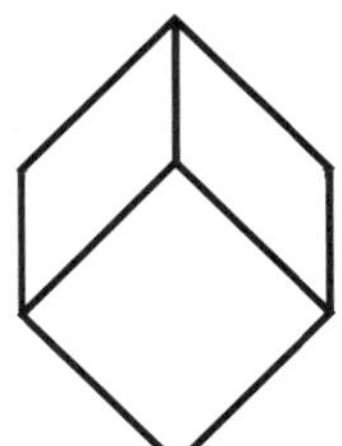 ______________

3 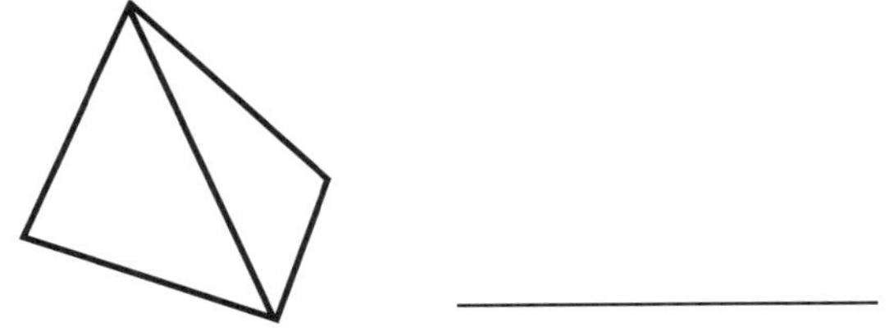 ______________

4  ______________

5 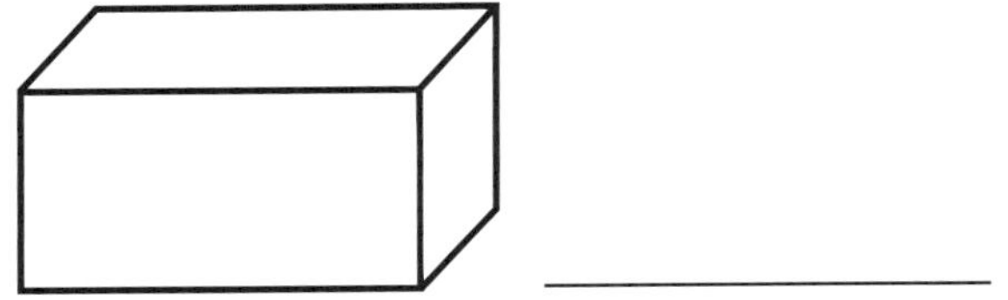 ______________

6 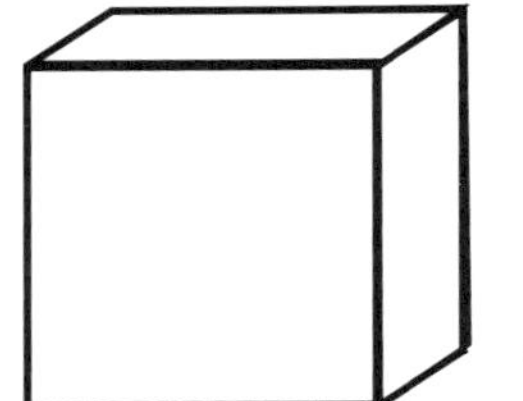 ______________

7 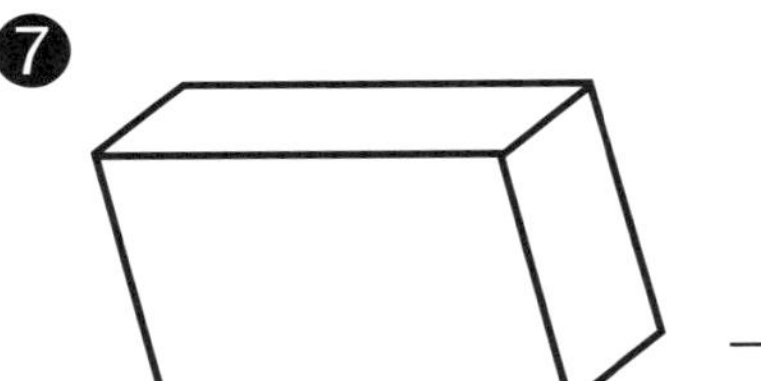 ______________

8 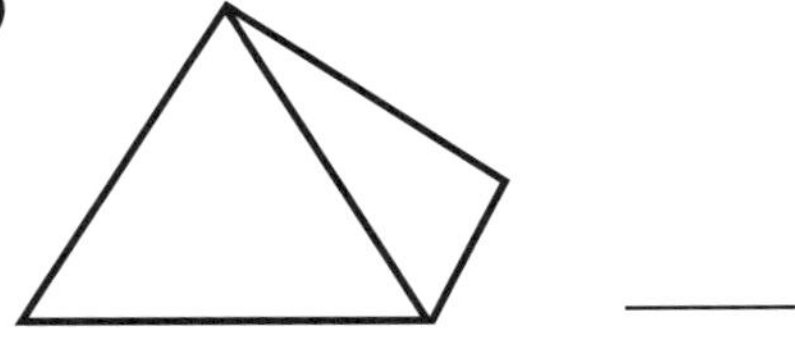 ______________

9 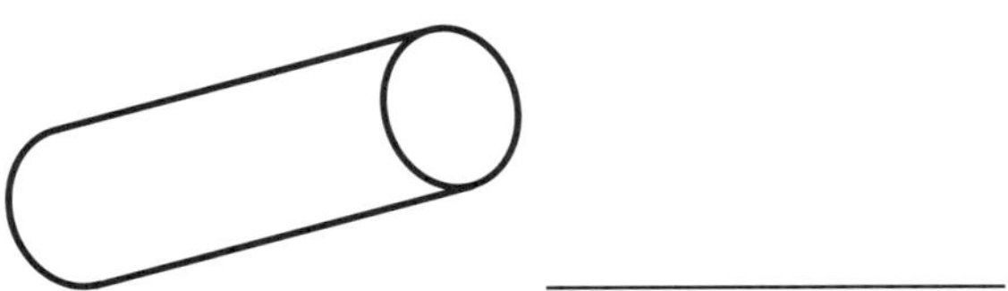 ______________

10 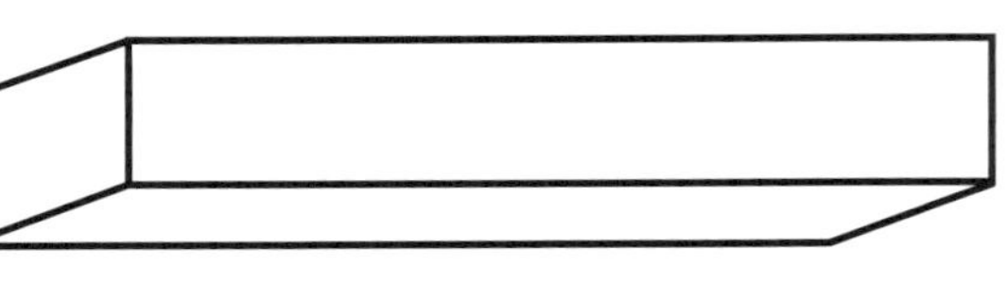 ______________

11 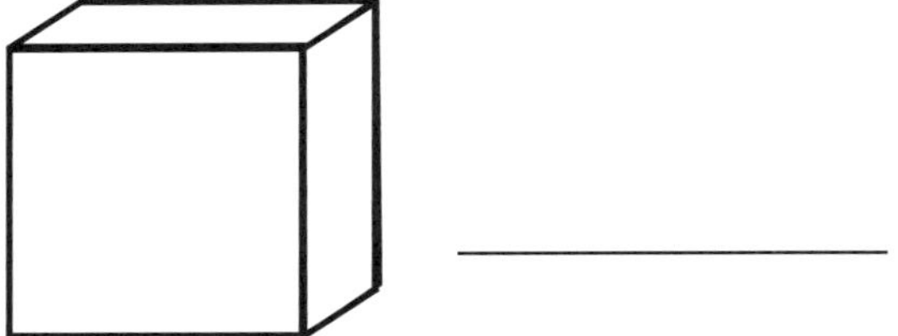 ______________

12 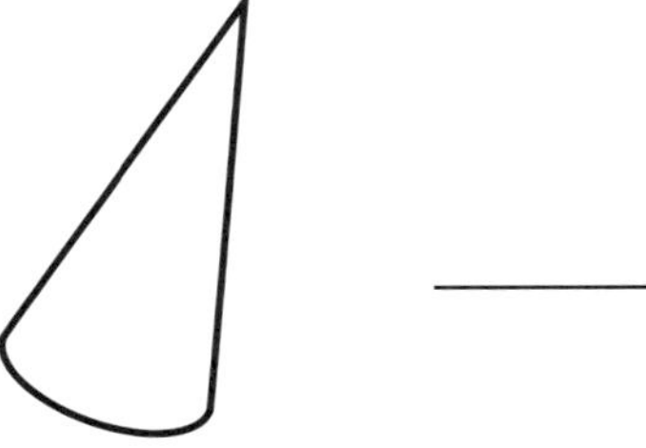 ______________

Name ______________________________ Date __________

Draw each 3-dimensional figure described. Write its name.

1. I have four equal sides and four equal angles. What am I?

____

2. I have six sides and six angles. What am I?

____

3. I have four sides and four angles. What am I?

____

4. I have six faces and six angles. What am I?

____

5. I have three sides and three angles. What am I?

____

6. I have five sides and five angles. What am I?

____

☆ **Tell how you know which shape is described.**

Unit 22 Mini-Lesson
Create New Figures

Standard

Geometry and Measurement

2.8E (SS) Decompose two-dimensional shapes, such as cutting out a square from a rectangle, dividing a shape in half, or partitioning a rectangle into identical triangles, and identify the resulting geometric parts.

Model the Skill

Hand out tangrams.

- **Say:** *We can use shapes to form or construct new shapes. We can also take apart or deconstruct shapes to form new shapes?* Guide students to use triangles to cover squares and rectangles. **Say:** *How many right triangles did you use to make a square?* (answers may vary) What about a rectangle?
- **Ask:** *How can you make a parallelogram*? (answers may vary)
- Assign students the appropriate practice page(s) to support their understanding of the skill. If necessary, have students fold another sheet of paper to see the equal shares for halves, fourths, and thirds.

Assess the Skill

Use the following problems to pre-/post-assess students' understanding of the skill.

Ask students to make smaller shapes by cutting the following shapes in halves, quarters, etc.

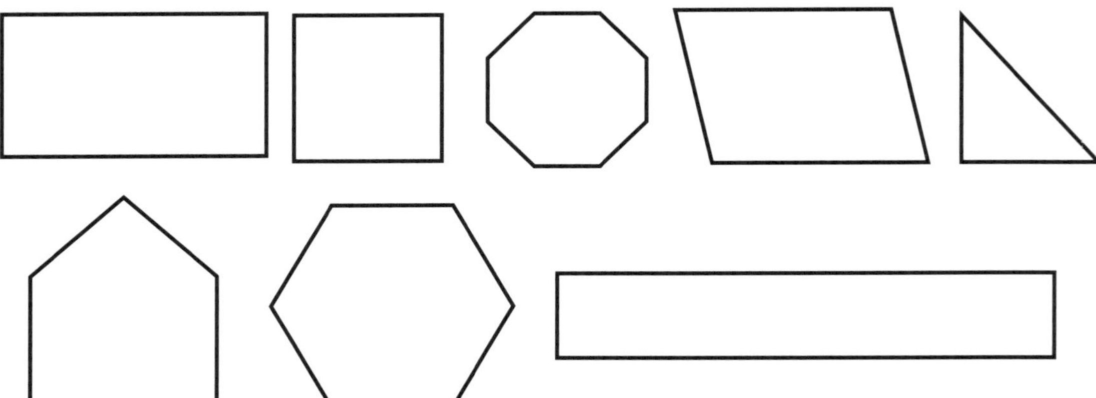

Name ______________________________ Date __________

Draw lines to make each shape into new shapes.

1. Make two rectangles.

2. Make three triangles.

3. Make one rectangle and four triangles.

4. Make four squares.

5. Make four triangles.

6. Make three triangles.

Tell how you make a square into four squares.

Name ______________________________ **Date** __________

Draw lines to make new shapes.

1. Make two rectangles and two triangles.

2. Make two squares and two triangles.

3. Make six rectangles.

4. Make nine squares.

5. Make one rectangle and four triangles.

6. Make four triangles.

☆ **Tell how you make a triangle into smaller triangles.**

Name ______________________________ **Date** __________

Solve.

 Make one rectangle and two triangles.

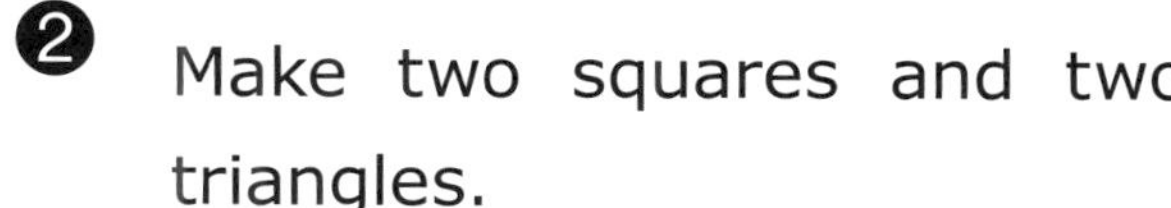 Make two squares and two triangles.

Circle the letter for the correct answer.

3. Which of the following shapes cannot be formed by combining all of the tangrams?

A triangle

B square

C pentagon

D parallelogram

 **Tell how two-fourths is like one-half.**

Unit 23 Mini-Lesson

Locate Points on a Number Line

Standard

Number and Operations

2.2E (SS) Locate the position of a given whole number on an open number line.

Model the Skill

Draw the following number line on the board.

- **Say:** *You can find different values on a number line. Each value has a different location. You can mark the location with a point.* Show students how to locate the values for 1, 2, and 3 on the number line and label those points. Then add mystery points *a, b,* and *c* as seen below.

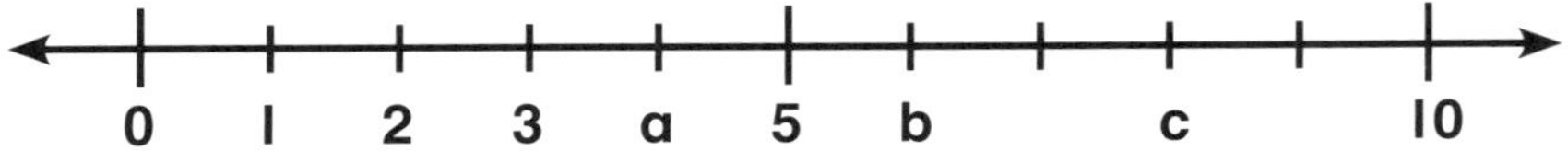

- **Ask:** *What is the value of point a on the number line?* (4) *What are the values of points b and c?* (6; 8)

- Assign students the appropriate activity page(s) to support their understanding of the skill. Remind students using number lines to circle the first addend and draw jumps to the right on the number line equal to the second addend.

Assess the Skill

Use the following problems to pre-/post-assess students' understanding of the skill.

Have students find the values for points d, e, and f.

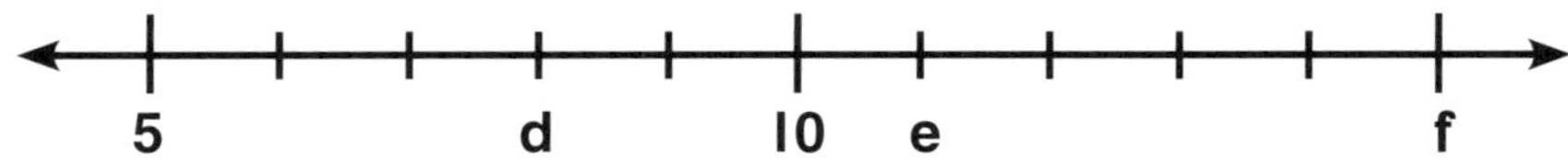

Name ______________________________ **Date** __________

Locate the point indicated on each number line.

1 **28**

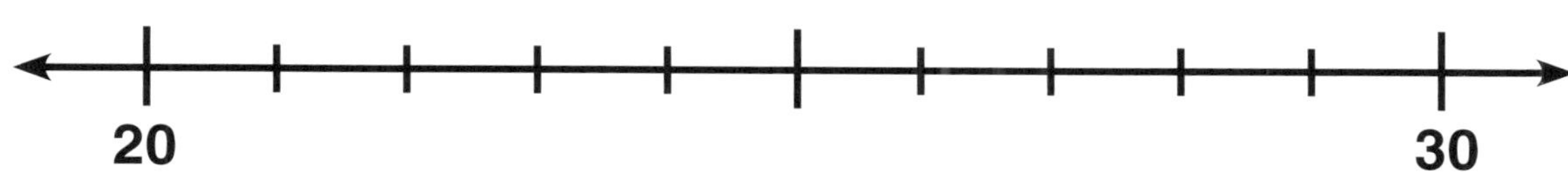

2 **31**

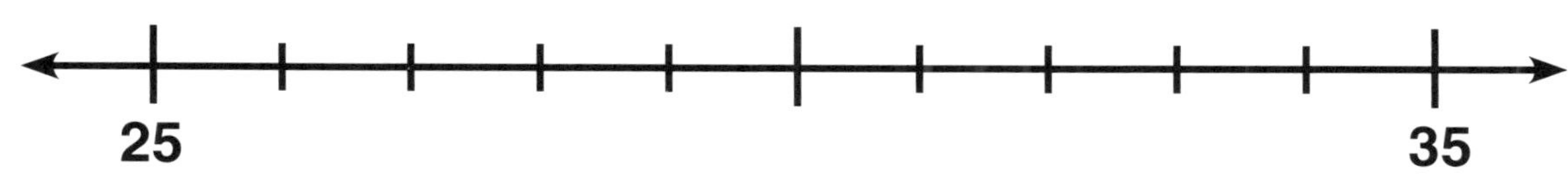

3 **29**

4 **50**

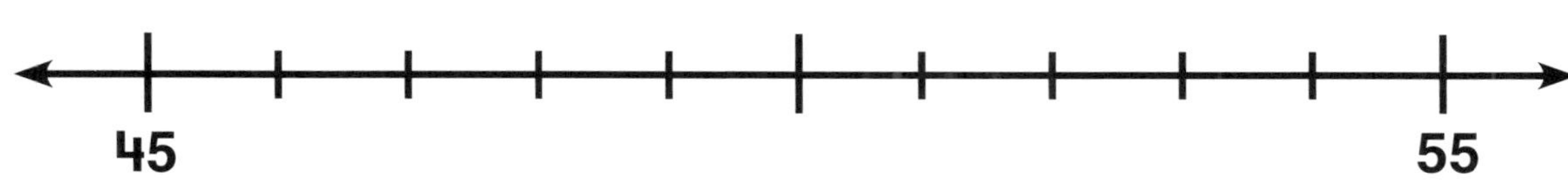

5 **63**

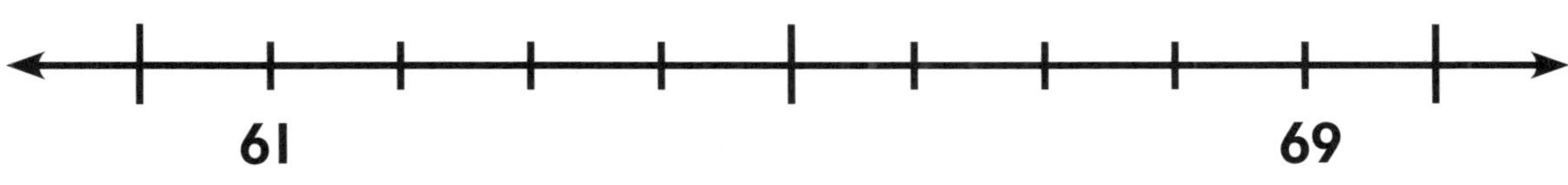

6 **76**

☆ **Tell how you found the point on the number line.**

Name ____________________ **Date** __________

Locate the point indicated on each number line.

❶ 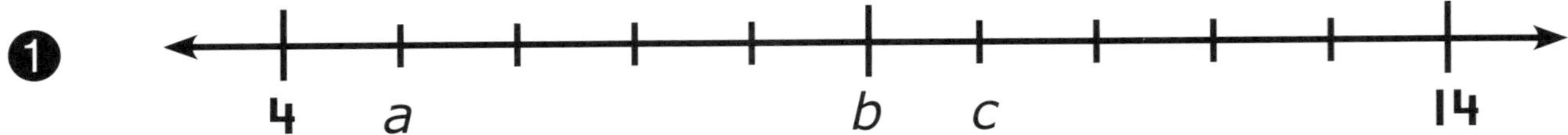

a =   b =   c =

❷ 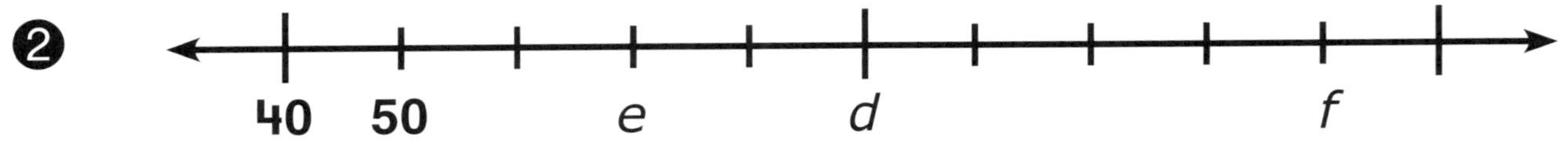

d =   e =   f =

❸ 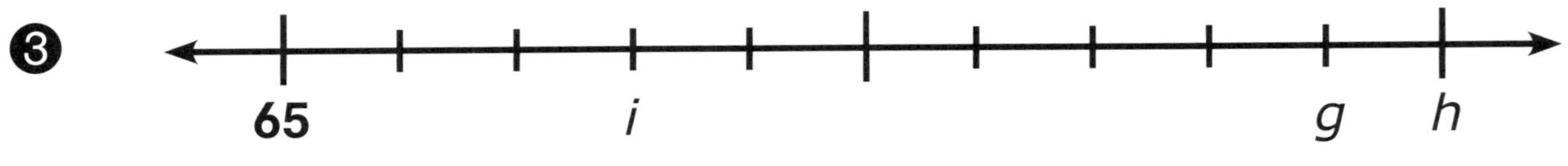

g =   h =   i =

❹ 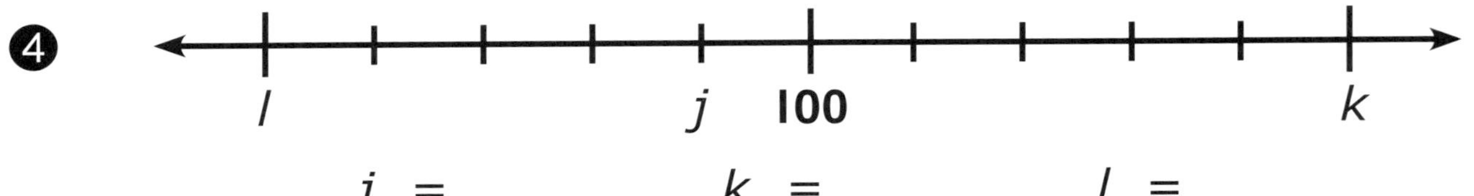

j =   k =   l =

❺ m   n   108   o

m =   n =   o =

❻ 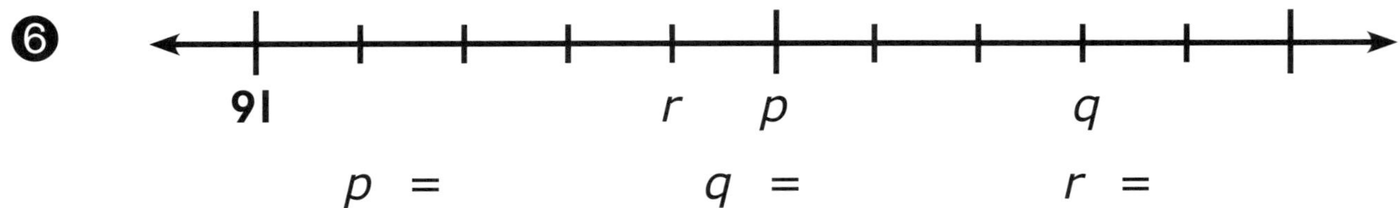

p =   q =   r =

☆ **Tell how you checked your answers.**

Name ______________________________ **Date** __________

Solve.

1. Point *A* has a value of 14.
Label point *A* on the number line.

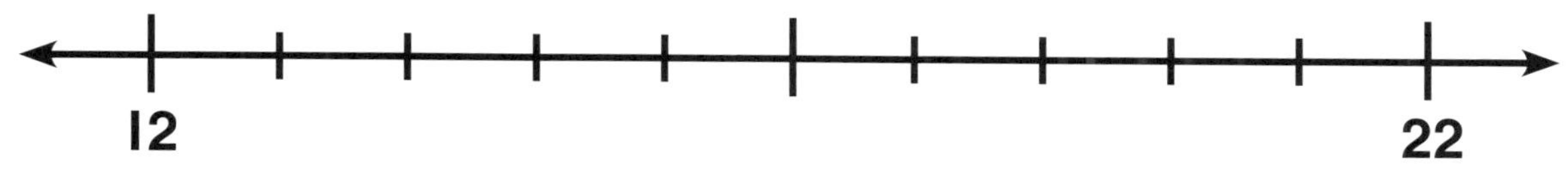

2. Point *C* has a value of 21.
Label point *C* on the number line.

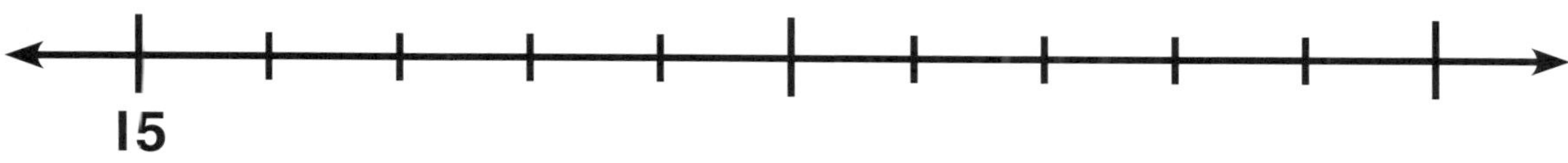

Choose the correct answer.

3. Which shows point *F* with a value of 41?

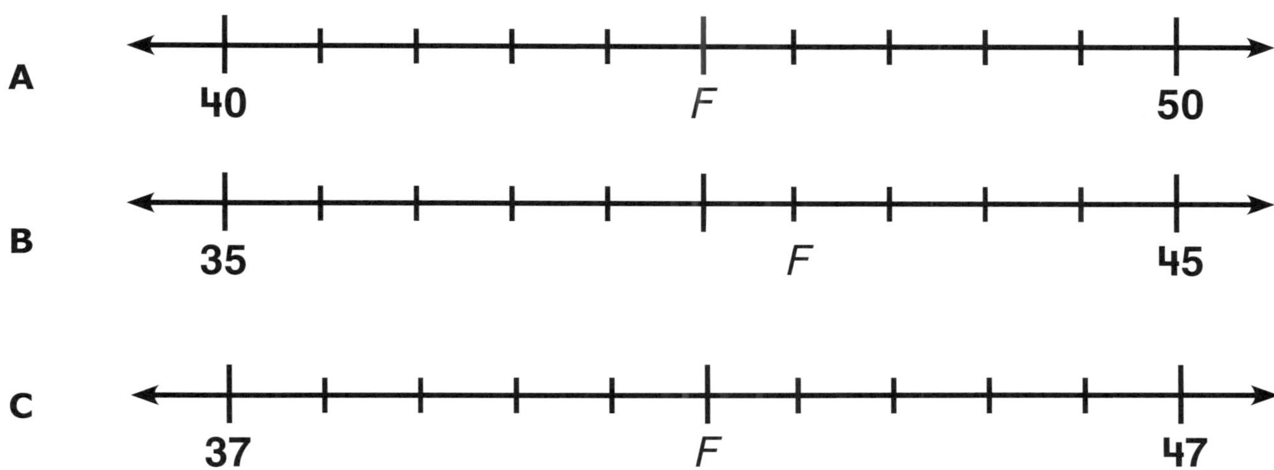

Unit 24 Mini-Lesson

Use Models for Inch and Foot

Standard

Geometry and Measurement

2.1C (PS) Select tools, including real objects, manipulatives, paper and pencil, and technology as appropriate, and techniques, including mental math, estimation, and number sense as appropriate, to solve problems.

2.9A (SS) Find the length of objects using concrete models for standard units of length.

Model the Skill

Hand out large paper clips, white sheets of paper, and an inch ruler.

- Display an inch ruler, a yardstick, and a tape measure, and have students identify each object. **Say:** *These are three different tools that can measure the length of an object. Look at your ruler. What units does a ruler use to measure?* (inches and feet) Discuss that a tape measure can also measure in inches and feet while a yardstick can measure in inches, feet, or yards.
- **Say:** *When you measure an object, line up the zero mark of the ruler with one edge of the object you are measuring. If there is no zero mark, line up the end of the ruler with the edge of the object. Look to see which inch measurement is closest to the other end of the object. About how long is a typical white sheet of paper?* (about 11 inches)
- Assign students the appropriate practice page(s) to support their understanding of the skill. Check that they line up the zero mark of the ruler with the left edge of the object and see which inch the other end of the object is closest to.

Assess the Skill

Use the following problems to pre-/post-assess students' understanding of the skill.

Have students use rulers to measure a variety of classroom items.

- Large paper clip—about 2 inches
- Ballpoint pen (with cap)—about 6 inches
- Stapler—about 7 inches
- Book or magazine—answers may vary

Name ________________________________ **Date** __________

Measure the length of the object.

❶ a desk

about ______ paper clips

about ______ pieces of paper

❷ door

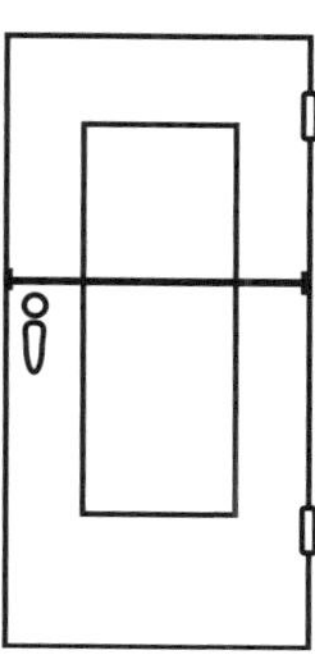

about ______ paper clips

about ______ pieces of paper

❸ a board

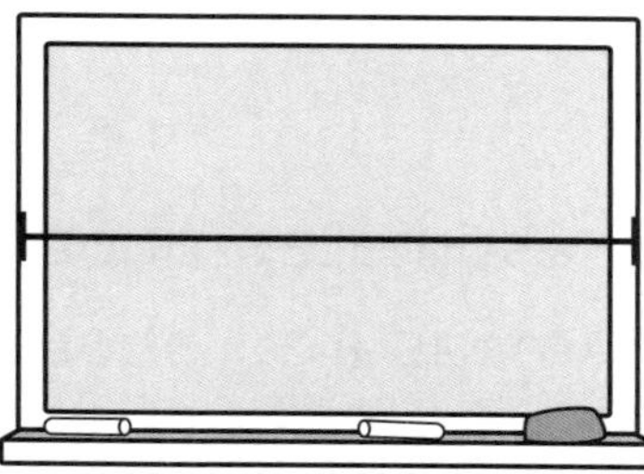

about ______ paper clips

about ______ pieces of paper

❹ a classmate

about ______ paper clips

about ______ pieces of paper

❺ a table

about ______ paper clips

about ______ pieces of paper

❻ a bookshelf

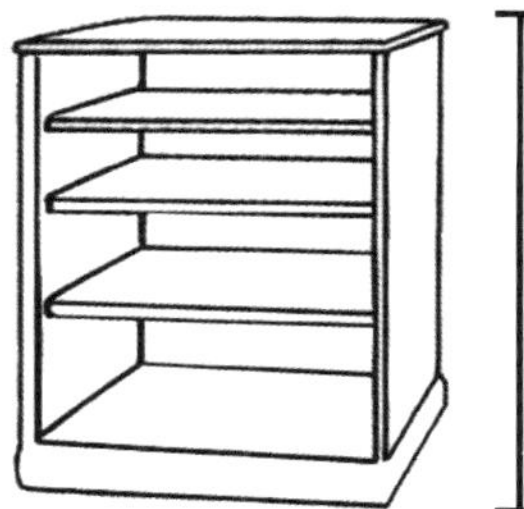

about ______ paper clips

about ______ pieces of paper

 **Tell how the two measurements relate to the size of the units.**

Name ______________________________ **Date** __________

Estimate the length of the object. Circle the unit. Then measure the object.

1 your pencil

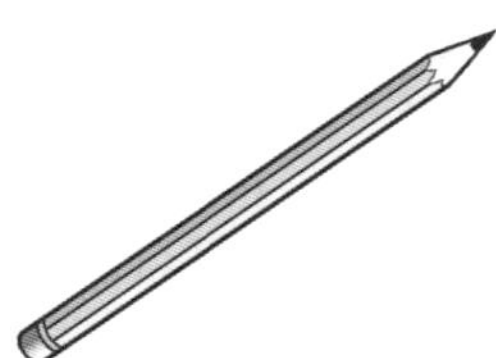

estimate: about _____ inches/feet
measure: about _____ inches/feet

2 a table

estimate: about _____ inches/feet
measure: about _____ inches/feet

3 a crayon

estimate: about _____ inches/feet
measure: about _____ inches/feet

4 a book

estimate: about _____ inches/feet
measure: about _____ inches/feet

5 a door

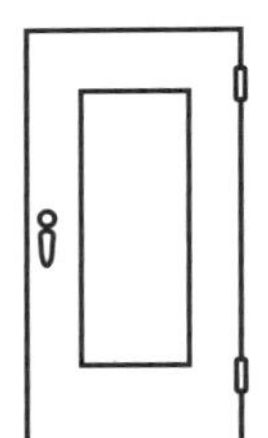

estimate: about _____ inches/feet
measure: about _____ inches/feet

6 a desk

estimate: about _____ inches/feet
measure: about _____ inches/feet

7 a classmate

estimate: about _____ inches/feet
measure: about _____ inches/feet

8 a window

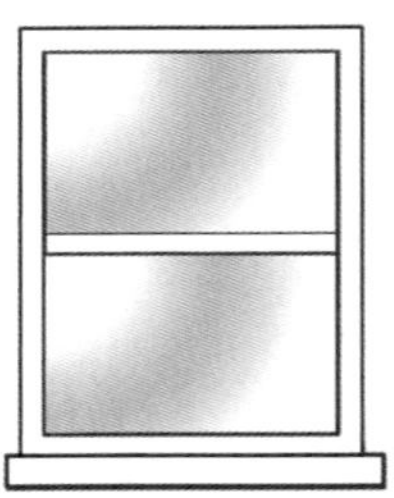

estimate: about _____ inches/feet
measure: about _____ inches/feet

☆ **Tell how you know your estimate is reasonable.**

Name ______________________________ **Date** __________

Measure the objects to compare their lengths. Circle the unit.

❶ How much longer is this activity page than a crayon?

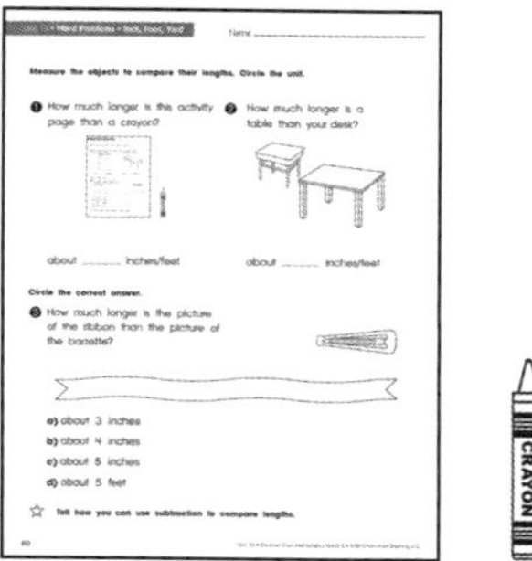
Measure the objects to compare their lengths. Circle the unit.

How much longer is this activity page than a crayon?

How much longer is a table than your desk?

about ______ inches/feet

about ______ inches/feet

Circle the correct answer.

How much longer is the picture of the ribbon than the picture of the barrette?

a) about 3 inches

b) about 4 inches

c) about 5 inches

d) about 5 feet

Tell how you can use subtraction to compare lengths.

about ________ inches/feet

❷ How much longer is a table than your desk?

about ________ inches/feet

Circle the correct answer.

❸ How much longer is the picture of the ribbon than the picture of the barrette?

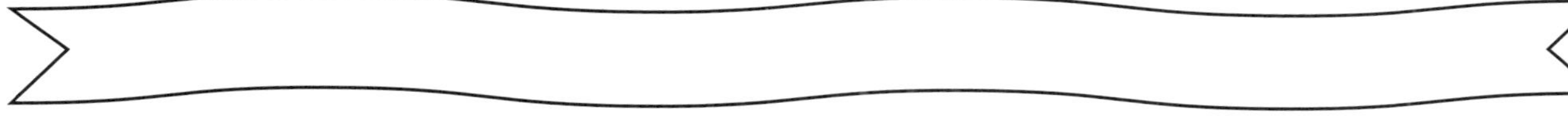

A about 3 inches

B about 4 inches

C about 5 inches

D about 5 feet

 **Tell how you can use subtraction to compare lengths.**

Unit 25 Mini-Lesson

Use Models for Centimeter and Meter

Standard

Geometry and Measurement

2.1C (PS) Select tools, including real objects, manipulatives, paper and pencil, and technology as appropriate, and techniques, including mental math, estimation, and number sense as appropriate, to solve problems.

2.9A (SS) Find the length of objects using concrete models for standard units of length.

Model the Skill

Hand out inch rulers.

- Display a centimeter ruler and a meter stick, and have students identify each object. **Say:** *These are three different tools that can measure the length of an object. Look at your ruler. What units does a ruler use to measure?* (millimeters and centimeters) Discuss that a meter stick can also measure in meters and centimeters.
- **Say:** *When you measure an object, line up the zero mark of the ruler with one edge of the object you are measuring. If there is no zero mark, line up the end of the ruler with the edge of the object. Look to see which millimeter measurement is closest to the other end of the object. About how long is a typical white sheet of paper?* (about 28 centimeters or 280 millimeters)
- Assign students the appropriate practice page(s) to support their understanding of the skill. Check that they line up the zero mark of the ruler with the left edge of the object and see which inch the other end of the object is closest to.

Assess the Skill

Use the following problems to pre-/post-assess students' understanding of the skill. Have students use rulers to measure a variety of classroom items.

- Large paper clip—about 5 centimeters
- Ballpoint pen (with cap)—about 15 centimeters
- Stapler—about 20 centimeters
- Book or magazine—about 25–30 centimeters

Name ______________________________ **Date** __________

Measure the object.

1. a table

about _______ erasers

about _______ folders

2. a window

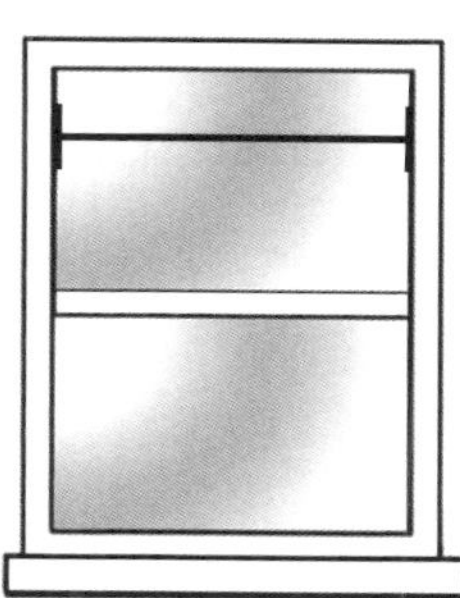

about _______ erasers

about _______ folders

3. a board

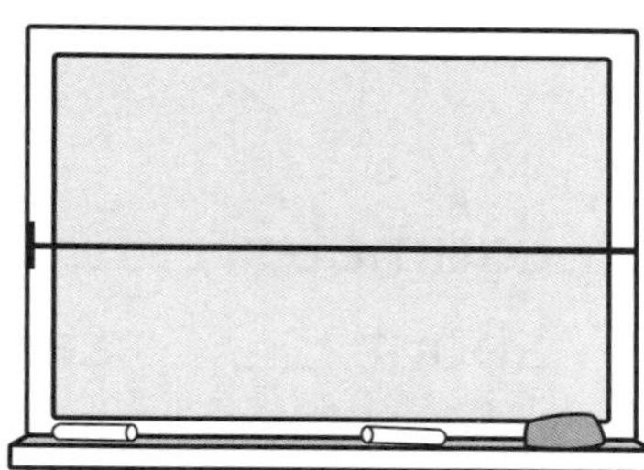

about _______ erasers

about _______ folders

4. a bookshelf

about _______ erasers

about _______ folders

5. a classmate

about _______ erasers

about _______ folders

6. the door

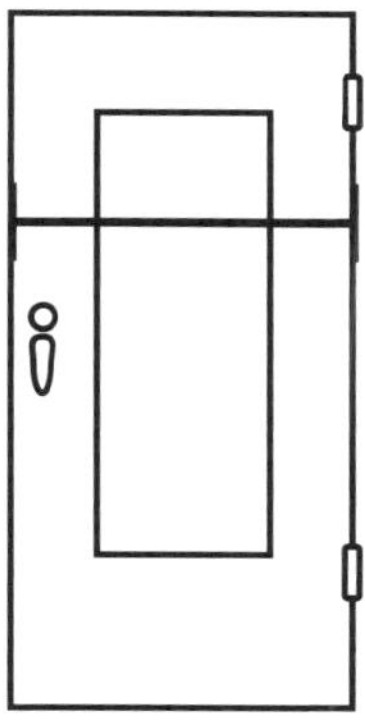

about _______ erasers

about _______ folders

☆ **Tell how the two measurements relate to the size of the units.**

Name ______________________________ **Date** __________

Estimate the length of the object. Circle the unit. Then measure the object.

1. a board eraser

estimate: ____ centimeters/meters
about ____ centimeters/meters

2. a board

estimate: ____ centimeters/meters
about ____ centimeters/meters

3. your shoe

estimate: ____ centimeters/meters
about ____ centimeters/meters

4. a friend

estimate: ____ centimeters/meters
about ____ centimeters/meters

5. a door

estimate: ____ centimeters/meters
about ____ centimeters/meters

6. a book

estimate: ____ centimeters/meters
about ____ centimeters/meters

7. the width of the hallway

estimate: ____ centimeters/meters
about ____ centimeters/meters

8. the length of the room

estimate: ____ centimeters/meters
about ____ centimeters/meters

☆ **Tell how you know your estimate is reasonable.**

Name ______________________________ **Date** ________

Measure the objects to compare their lengths. Circle the unit.

1. How much longer is this activity page than a crayon?

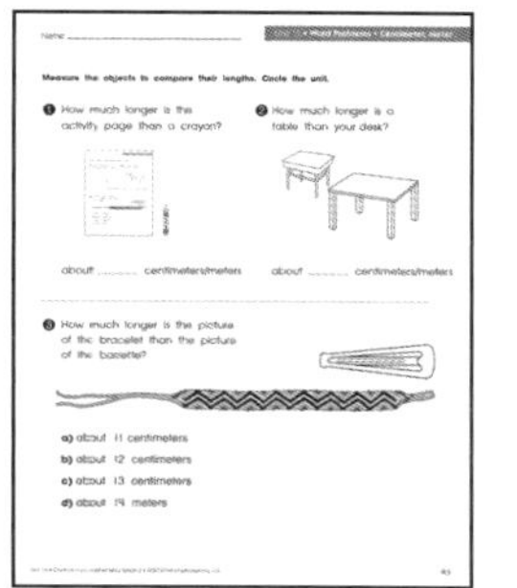

about _____ centimeters/meters

2. How much longer is a table than your desk?

about _____ centimeters/meters

---

3. How much longer is the picture of the bracelet than the picture of the barrette?

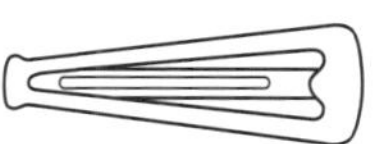

A about 11 centimeters

B about 14 centimeters

C about 13 centimeters

D about 14 meters

Unit 26 Mini-Lesson

Use Non-Standard Units to Determine Area

Standard

Geometry and Measurement

2.9F (SS) Use concrete models of square units to find the area of a rectangle by covering it with no gaps or overlaps, counting to find the total number of square units, and describing the measurement using a number and the unit.

Model the Skill

Draw the following figures on the board.

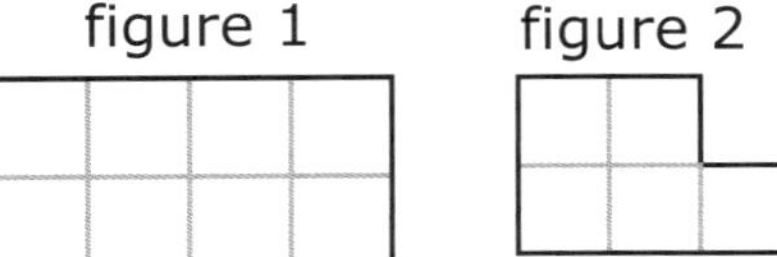

- **Say:** *Today we are going to learn about area. Area is the number of square units that are needed to cover a flat surface.* Have students look at the first figure and identify one square unit. **Ask:** *What figure do the square units make?* (rectangle) *You can count the square units in the rectangle to find its area. How many square units are in the rectangle?* (8) *We can say that the area of this rectangle is 8 square units.*
- Have students look at the second figure. **Ask:** *How is this figure different from figure 1?* (Possible response: This is not a rectangle; this figure has more sides, fewer square units.) Point out that students can still tell the area of the figure by counting the number of square units. (5) Allow students to use square tiles if they wish to model the figure and then find the area.
- Assign students the appropriate practice page(s) to support their understanding of the skill.

Assess the Skill

Use the following problems to pre-/post-assess students' understanding of the skill.

Say: *Find the area of a tabletop, desktop, or a piece of paper. Draw a closed shape with an area of 18 inches.*

Name ______________________________ Date __________

Count the square units to find the area.

1

_______ square units

2

_______ square units

3

_______ square units

4

_______ square units

5

_______ square units

6

_______ square units

Tell how you got your answer.

Name ______________________________ **Date** __________

Add to find the area.

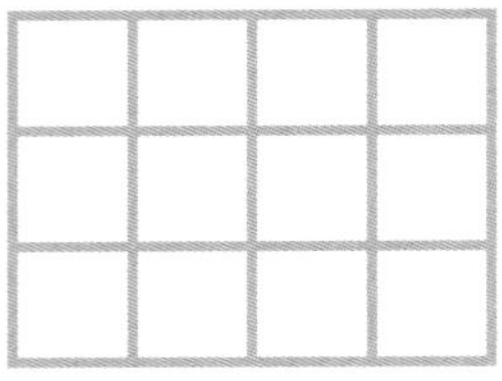

4 squares in each row, 3 rows

4 + 4 + 4 = ________

Area: ________ square units

2

___ squares in each row, ___ rows

5 + 5 = ________

Area: ________ square units

3

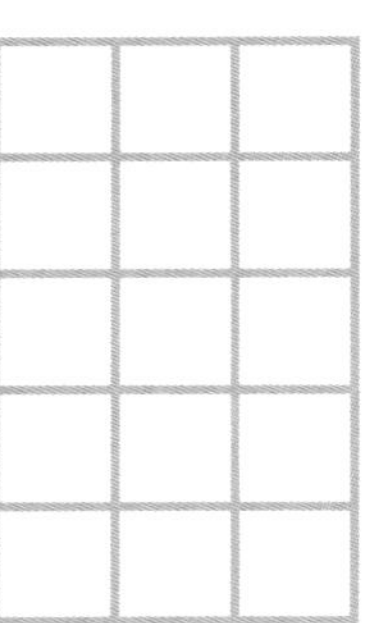

____

Area: ________ square units

4

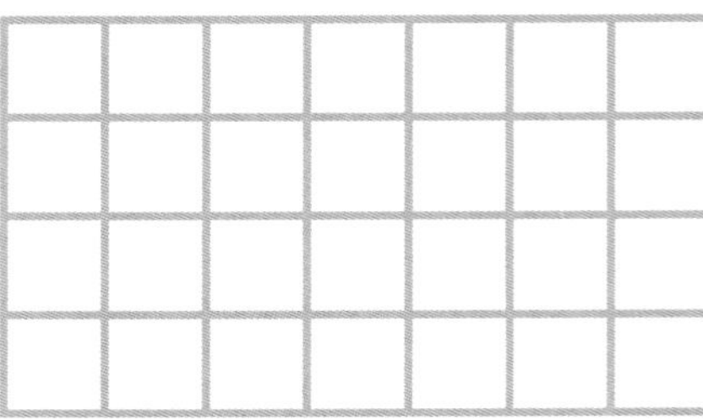

____

Area: ________ square units

5

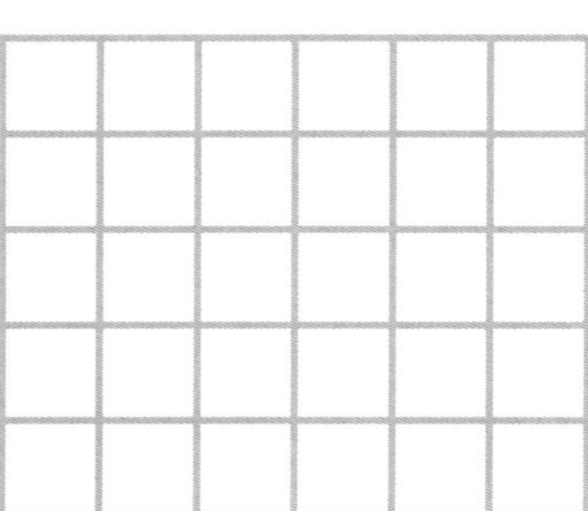

____

Area: ________ square units

6

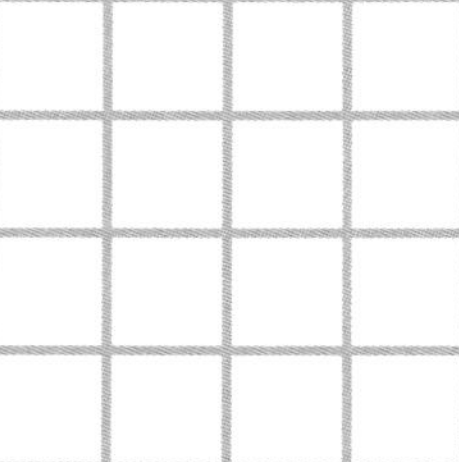

____

Area: ________ square units

 **Tell another addition sentence you can use to find the area.**

Name ______________________________ **Date** __________

Solve.

1. A rectangle has a length of 6 units and a width of 4 units. What is the area of the rectangle?

2. A square has a length of 5 units. What is the area of the square?

3. The quilt is a rectangle with a length of 10 square patches and a width of 8 square patches. What is the area of the quilt?

4. The tray has a length of 5 square tiles and a width of 3 square tiles. What is the area of the tray?

5. The gym floor has a length of 30 square floor tiles and a width of 20 floor tiles. What is the area of the gym?

   A 60 square floor tiles
   B 100 square floor tiles
   C 50 square floor tiles
   D 600 square floor tiles

6. Esther's rectangular patio is 9 tiles long and 7 tiles wide. If Esther covers the entire patio in tiles, how many tiles will she need?

   A 16 tiles
   B 32 tiles
   C 63 tiles
   D 126 tiles

Unit 27 Mini-Lesson

Use Non-Standard Units to Determine Capacity and Weight

Standard

Mathematical Process Standards

2.1C (PS) Select tools, including real objects, manipulatives, paper and pencil, and technology as appropriate, and techniques, including mental math, estimation, and number sense as appropriate, to solve problems.

Model the Skill

Hand out identical objects, such as pencils or nickels to each student.

- **Say:** *Today we are going to learn about mass. The mass of an object tells the amount of matter in that object. On Earth, it also tells us how heavy an object is. You can measure mass in grams or kilograms. One kilogram is equal in mass to 1,000 grams.* Display a balance and some gram weights. **Say:** *I could use grams and a balance to measure the mass of a crayon or an eraser. What other objects might I measure this way?* (Possible response: small, light objects such as paper or a pen) Allow students to brainstorm other objects that might be measured using grams and a balance, and other objects that might be measured using kilograms and a scale.
- Have students work together to estimate and then measure their nickels. To help students estimate, have them hold the weights and the object they will measure. Remind students that an estimate is a thoughtful guess.
- **Ask:** *Would it be better to use grams or kilograms to measure the mass of a cat? Why?* (Possible response: kilograms because grams are too small a unit of measurement) Have students suggest other items that might be measured better in grams or kilograms.
- Assign students the appropriate practice page(s) to support their understanding of the skill.

Assess the Skill

Use the following problems to pre-/post-assess students' understanding of the skill.

Ask students to estimate and then find the mass of the following objects:
- a ruler
- a pen
- a shoe
- a full bookbag or backpack

Name ____ **Date** __________

Circle the better estimate of the mass of each item pictured.

bananas

1 gram

1 kilogram

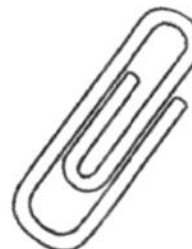

paper clip

1 gram

1 kilogram

dog

10 grams

10 kilograms

❹

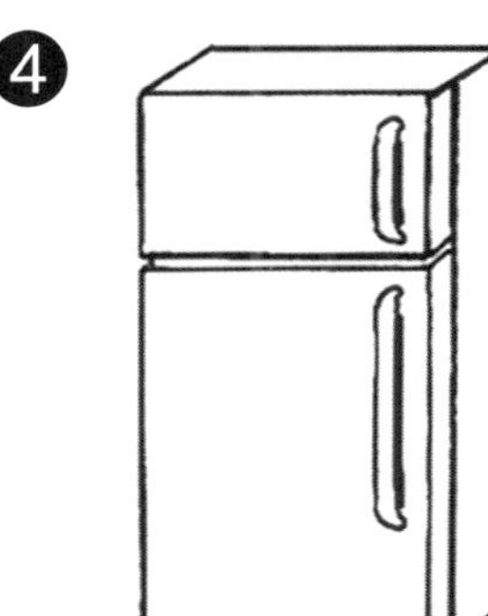

refrigerator

200 grams

200 kilograms

Circle the better estimate of the capacity of each item pictured.

water bottle

1 liter

10 liters

birdbath

1 liter

5 liters

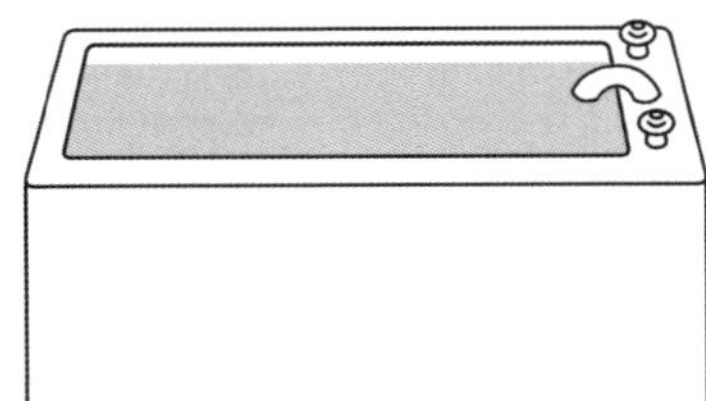

bathtub

5 liters

50 liters

❽

saucepan

3 liters

30 liters

Measure the mass of a paper clip. Tell how you know your answer is reasonable.

Name ______________________________ **Date** __________

Estimate the mass and capacity of each container. Then measure.

1 **Mass**

Object	**Estimate**	**Measure**
teaspoon		
paper cup		
large bottle		

2 **Capacity**

Object	**Estimate** **More or less than a liter?**	**Measure** **More or less than a liter?**
teaspoon		
paper cup		
large bottle		

 **Tell something that has a capacity of about 1 liter.**

Name ____ **Date** __________

How much will each container hold? Circle the better estimate.

drinking glass
more than a liter
less than a liter

soup can
more than a liter
less than a liter

3

pitcher
3 liters
30 liters

fish tank
2 liters
20 liters

Circle the correct answer for each question.

5 Olivia fills the bathtub full of water. About how much water is in the bathtub?

A 40 grams
B 4 kilograms
C 4 liters
D 40 liters

6 Rex has a toaster in his kitchen. What is the approximate mass of his toaster?

A 10 grams
B 1 kilogram
C 1 liter
D 10 liters

Unit 28 Mini-Lesson

Read a Thermometer

Standard

Mathematical Process Standards

2.1C (PS) Select tools, including real objects, manipulatives, paper and pencil, and technology as appropriate, and techniques, including mental math, estimation, and number sense as appropriate, to solve problems.

Model the Skill

Show examples of a thermometer including these drawn here.

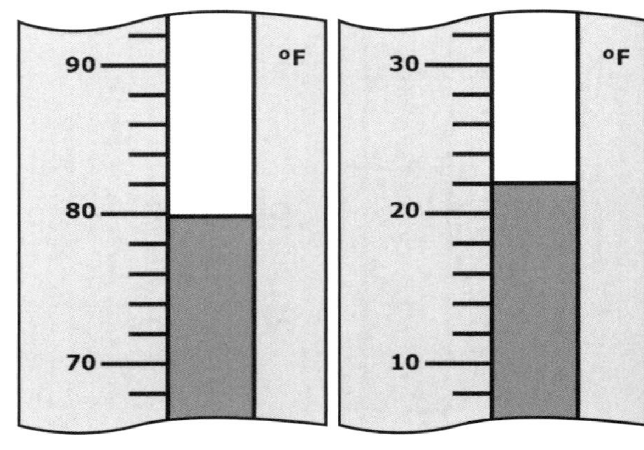

Thermometer A **Thermometer B**

- **Say:** *One way we measure temperature is by using a thermometer. These thermometers measure temperature on a scale of degrees Fahrenheit. A thermometer is a lot like a number line. We can use our understanding of number lines to read a thermometer.*
  On this thermometer, each longer tick mark is 10 degrees. Each short tick mark is 2 degrees.

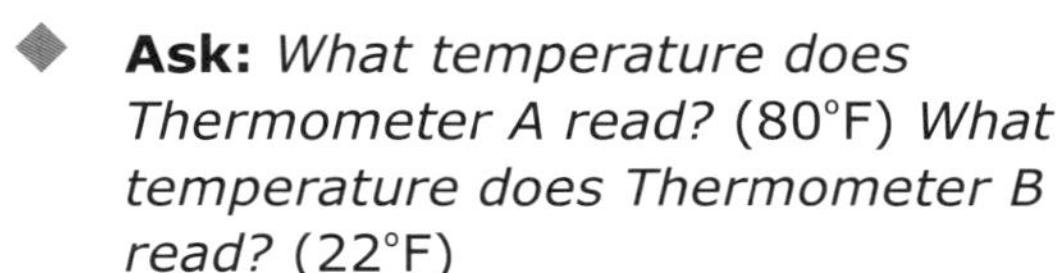

- **Ask:** *What temperature does Thermometer A read?* (80°F) *What temperature does Thermometer B read?* (22°F)

- Assign students the appropriate practice page(s) to support their understanding of the skill.

Assess the Skill

Use the following problems to pre-/post-assess students' understanding of the skill.

Ask students to identify the thermometers that show the warmest and coolest temperatures.

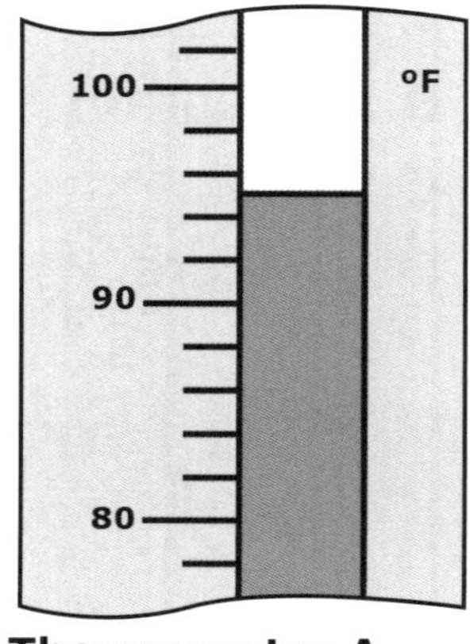

Thermometer A

70
60
50
°F

Thermometer B

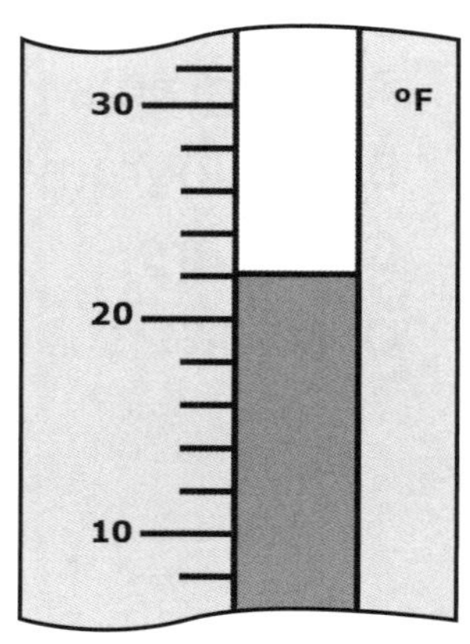

Thermometer C

Name ________________________________ **Date** __________

Write the correct temperature for each thermometer.

❶

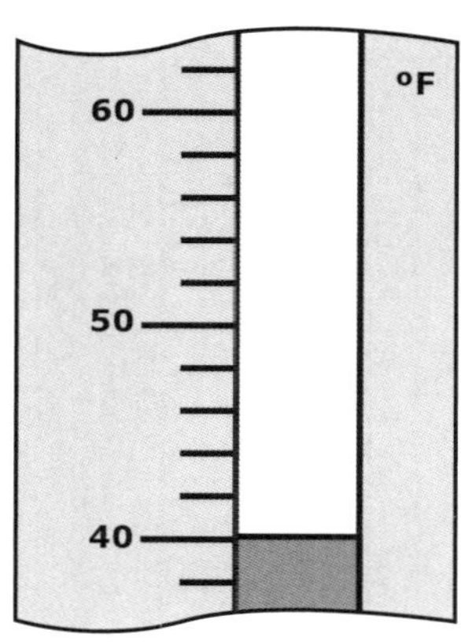

____

❷

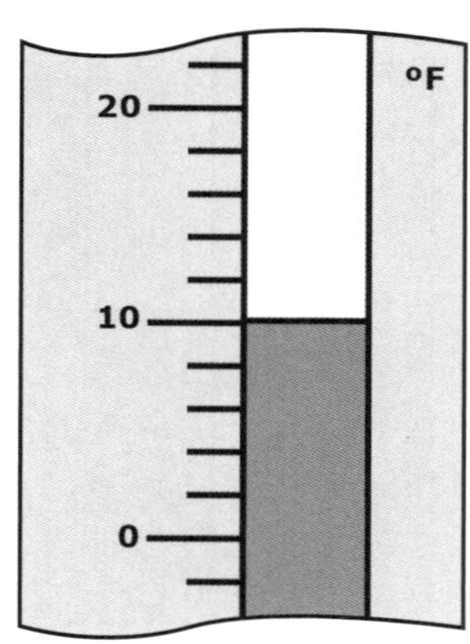

____

❸

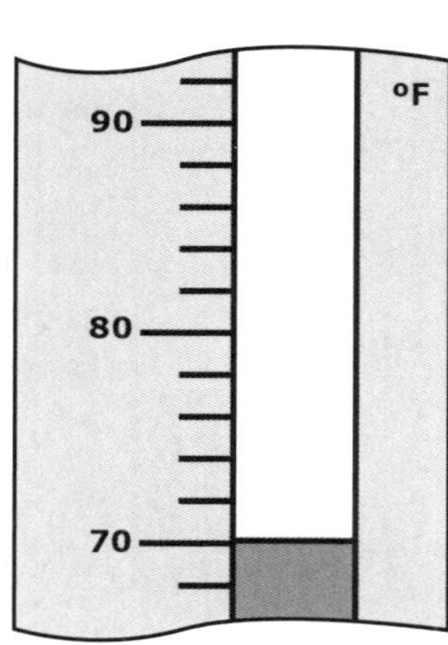

____

❹

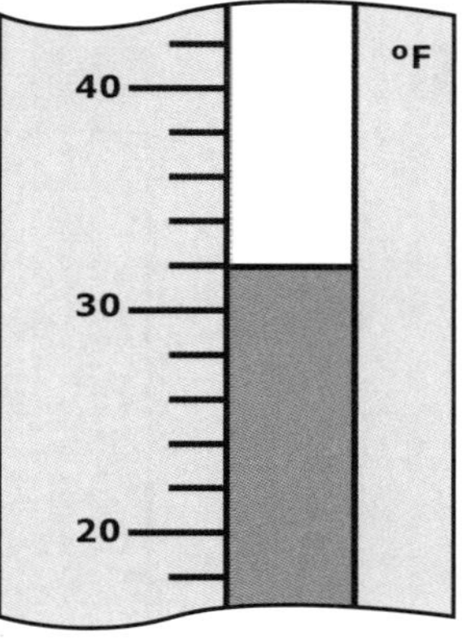

____

❺

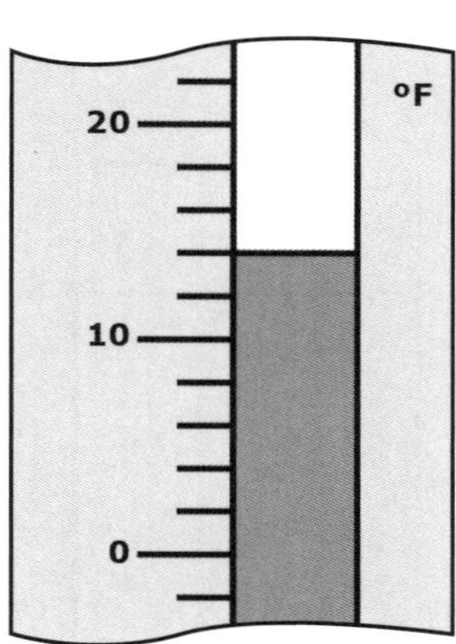

____

❻

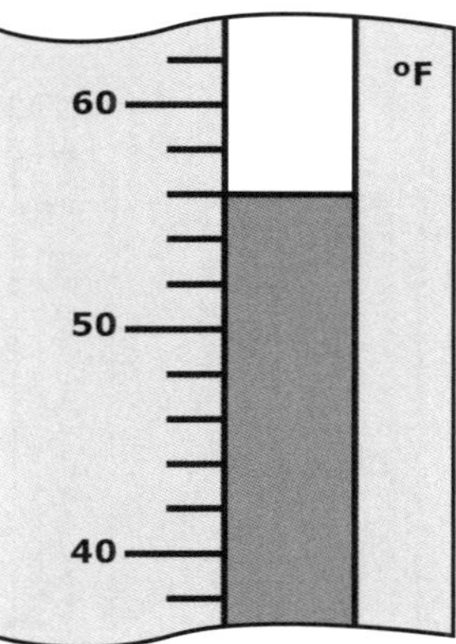

____

☆ **Tell how you found the temperature.**

Name ______________________________ **Date** __________

Write the correct temperature for each thermometer.

❶
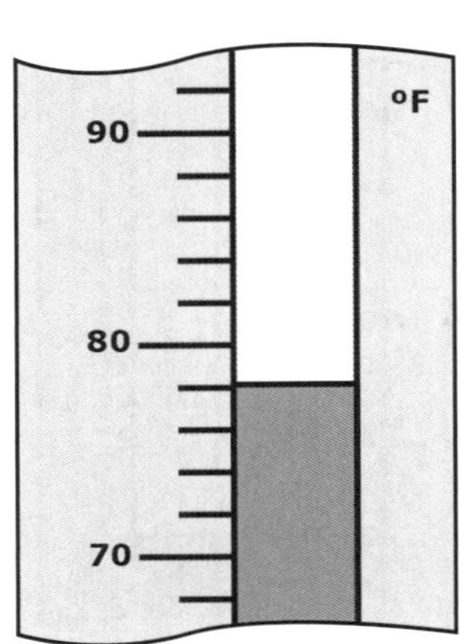

____

❷
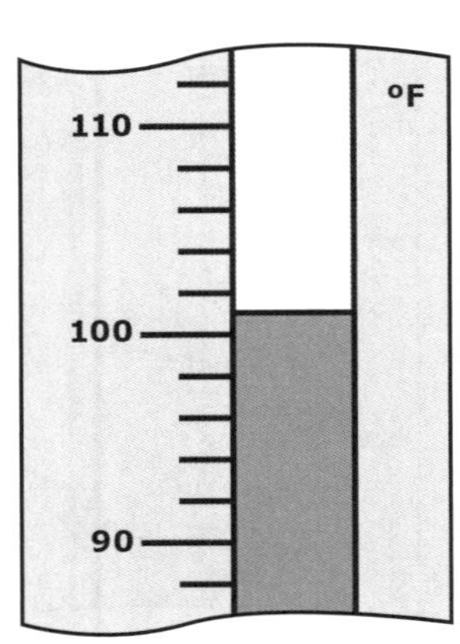

____

❸
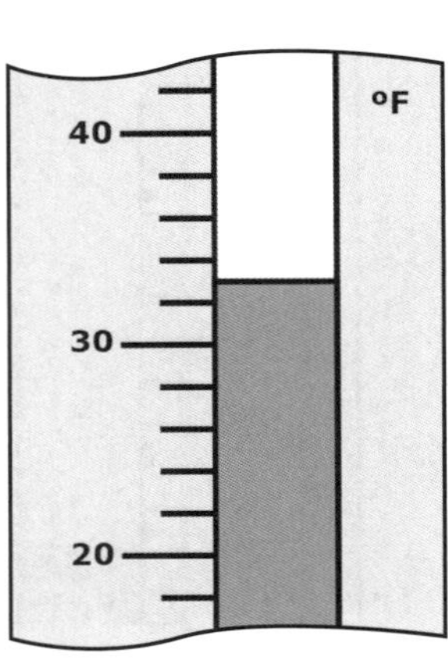

____

Shade each thermometer to show the temperature indicated.

❹
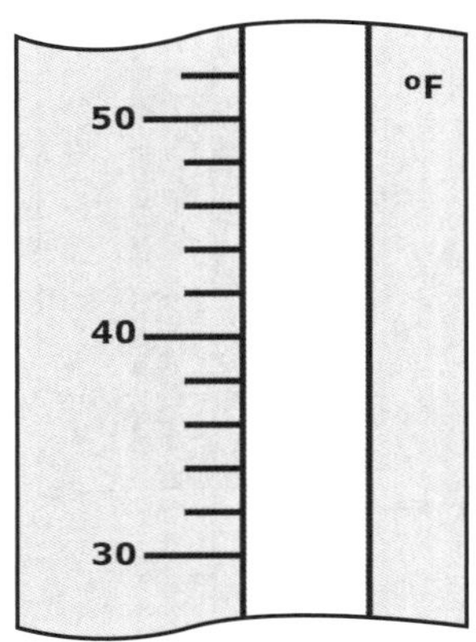

42 degrees

❺
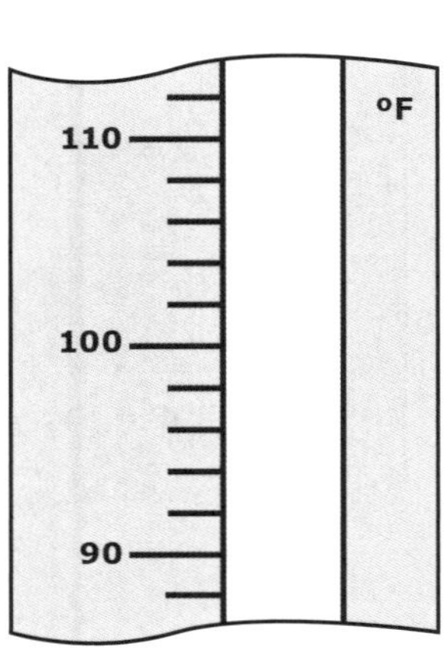

98 degrees

❻
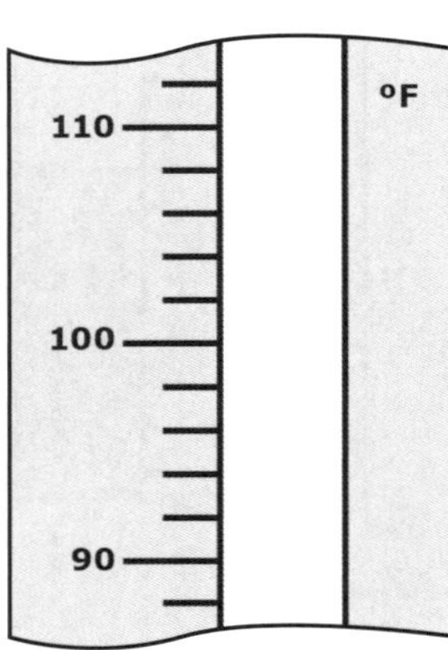

103 degrees

☆ **Tell how you found the number of degrees for each tick mark.**

Name ____ Date __________

1. What temperature is shown on the thermometer below?

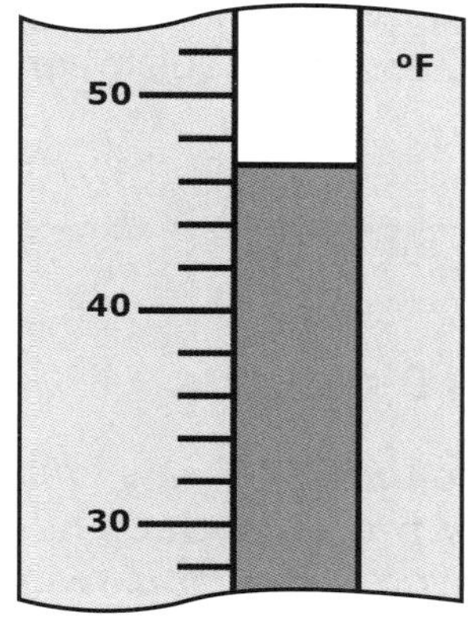

____

2. Label and shade the thermometer below to show a temperature less than 100 degrees.

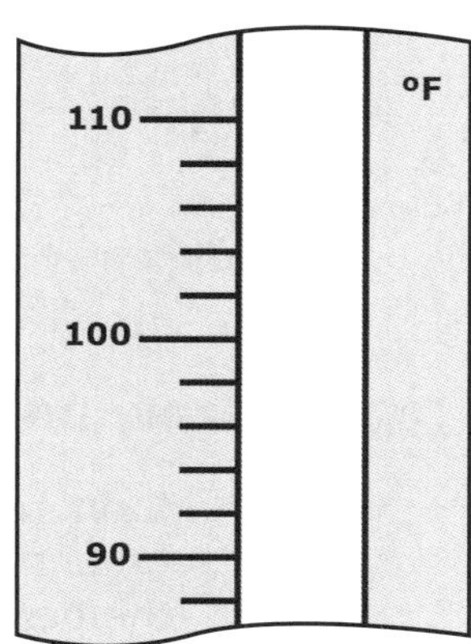

____

Circle the letter for the correct answer.

3. Which of the thermometers below shows a temperature greater than 110 degrees?

A

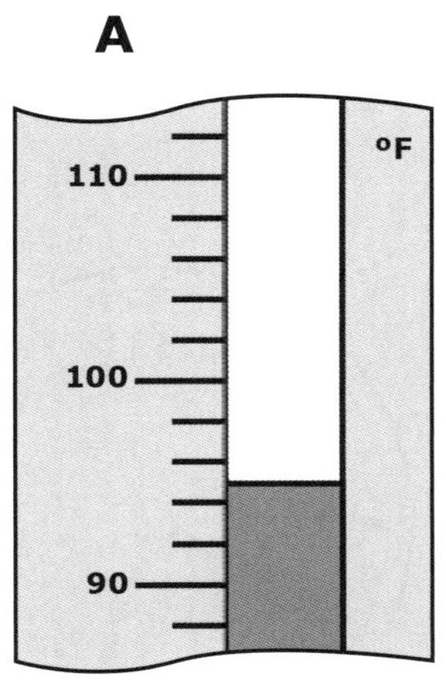

B

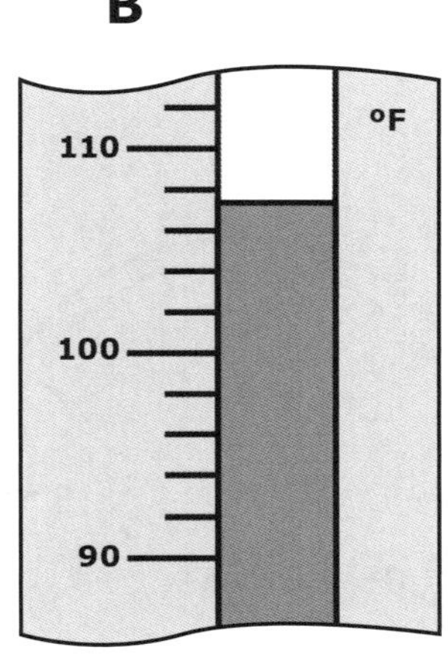

C

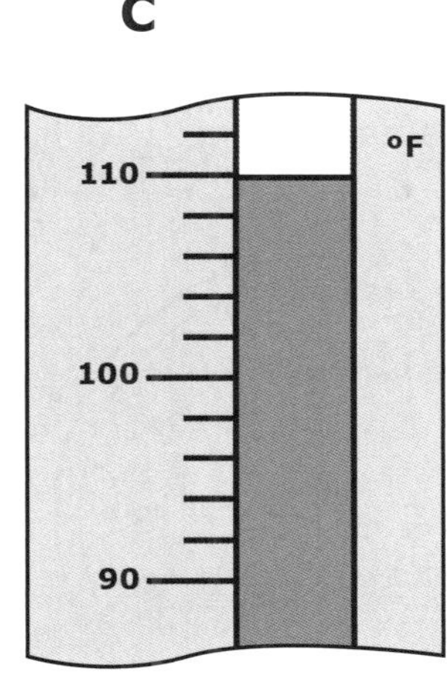

D

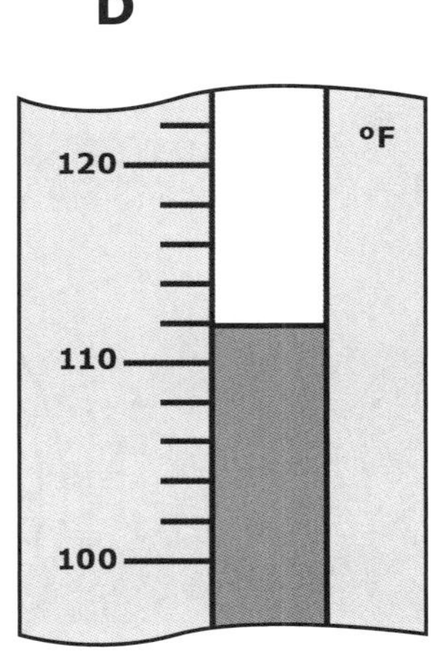

Unit 29 Mini-Lesson

Tell Time to the Nearest Minute

Standard

Geometry and Measurement

2.9G (RS) Read and write time to the nearest one-minute increment using analog and digital clocks, and distinguish between a.m. and p.m.

Model the Skill

Draw an analog clock that shows 10 o'clock.

- **Say:** *You can show the same time in more than one way. What time is shown on this clock?* (10 o'clock) Write **10 o'clock** on the board. **Say:** *There is another way to write* ***10 o'clock*** *with only numbers.* Write **10:00**. **Say:** *The numbers before the colon tell the hour and the numbers after the colon tell the minutes.* Point out that 10:00 says hour 10 and zero minutes.

Draw an analog clock that shows 10:15.

- **Say:** *Look at this clock. What time is shown on the digital clock?* (10:15, or a quarter past ten) **Ask:** *How do you know that it is 10:15?*
- Assign students the appropriate practice page(s) to support their understanding of the skill.

Assess the Skill

Use the following problems to pre-/post-assess students' understanding of the skill.

Write each time.

Name ______________________________ **Date** __________

Draw lines to match clocks that show the same times.

1

2

3

4

5

6

7

8

10:39
1:00
1:31
8:55
12:05
6:48
2:03
4:30

☆ **Tell how you know when it is five minutes past the hour.**

Name ______________________________ **Date** __________

Write each time.

  ______ : ______

2  ______ : ______

3  ______ : ______

4  ______ : ______

5  ______ : ______

6  ______ : ______

7  ______ : ______

8  ______ : ______

9  ______ : ______

10  ______ : ______

11  ______ : ______

12  ______ : ______

 **Tell how you tell time to the nearest five minutes.**

Name ____ **Date** __________

Solve.

1. Pipa left for school at 8:25 A.M.
Circle the clock that shows which time she left.

2. David has practice at 3:07 P.M. Circle the clock that shows the time of practice.

Choose the correct answer for each problem.

3. What time is shown below?

A 7:35 A.M.
B 7:45 P.M.
C 7:45 A.M.
D 9:35 A.M.

4. What time is shown below?

A 5:43 A.M.
B 5:43 P.M.
C 8:29 A.M.
D 8:29 P.M.

Unit 30 Mini-Lesson

Estimate How Much Time It Takes

Standard

Mathematical Process Standards

2.1F (PS) Analyze mathematical relationships to connect and communicate mathematical ideas.

Model the Skill

Draw an analog clock that shows 6 o'clock.

- **Say:** *Different activities we perform each day take different amounts of time to complete. Last night I went to bed at around 10 o'clock. This morning I woke up at around 6 o'clock. So about how long did I sleep for? Several seconds? Several minutes? Or several hours?* (about 8 hours) *The first thing I did was wash my face. About how much time does that take?* About 45 seconds? 45 minutes? Or 45 hours? (45 seconds)
- **Say:** *I started my breakfast at about 6:15.* **Ask:** *How long do you think it took me to prepare and eat my breakfast? 15 seconds, minutes, or hours?* (15 minutes)
- Assign students the appropriate practice page(s) to support their understanding of the skill.

Assess the Skill

Use the following problems to pre-/post-assess students' understanding of the skill.

- Ask students to give examples of activities that take approximately
  - one second
  - one minute
  - one hour
  - one day
  - one week

Name ______________________________ **Date** __________

Estimate how long it takes to do each task. Circle the correct answer.

1. Walk dog
   12 seconds
   12 minutes
   12 hours

2. Brush teeth
   2 seconds
   2 minutes
   2 hours

3. Drive to school
   15 seconds
   15 minutes
   15 hours

4. Bake a cake
   45 seconds
   45 minutes
   45 hours

5. Grow an inch
   1 day
   1 week
   1 year

6. Grow a flower from a seed
   3 hours
   3 days
   3 months

Name ______________________________ **Date** __________

1. **Describe an activity that takes one second.**

2. **Describe an activity that takes one minute.**

3. **Describe an activity that takes one hour.**

4. **Describe an activity that takes one day.**

5. **Describe an activity that takes one week.**

6. **Describe an activity that takes one month.**

7. **Describe an activity that takes one year.**

Name ______________________________ **Date** __________

Solve.

1. It is 1 o'clock. If Helena has to make a sandwich and eat lunch, at what time will she be done? Circle the clock that shows the right time.

2. David has practice at 3:15 P.M. Circle the clock that shows the time practice will probably end.

Choose the correct answer for each problem.

3. About how long will it take to walk onto the bus?

A 10 seconds
B 10 minutes
C 10 hours
D 10 days

4. About how long does it take to read a story before bed?

A 15 seconds
B 15 minutes
C 15 hours
D 15 days

Unit 31 Mini-Lesson

Make a Line Plot

Standard

Data Analysis

2.10B (RS) Organize a collection of data with up to four categories using pictographs and bar graphs with intervals of one or more.

Model the Skill

Draw this tally chart and corresponding line plot on the board.

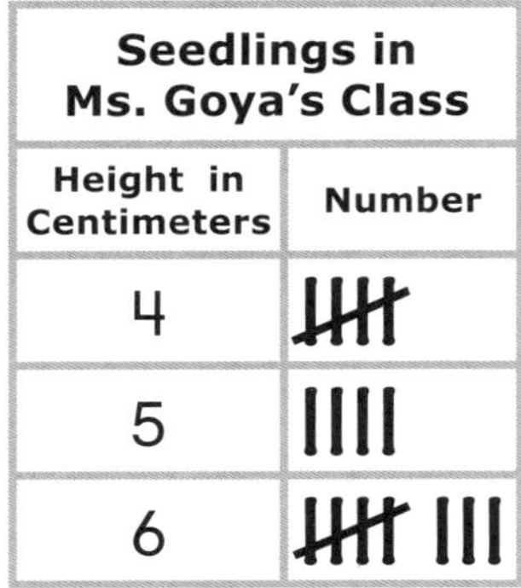

Seedlings in Ms. Goya's Class	
Height in Centimeters	**Number**
4	𝍸
5	||||
6	𝍸 |||

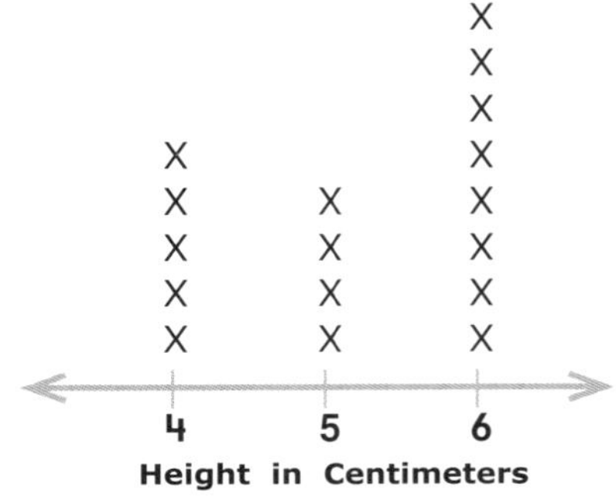

- **Say:** *Data can be shown in charts and graphs. A line plot uses* ***Xs*** *to show data on a number line. What does this line plot show?* (the heights of the seedlings in Ms. Goya's class) *What are the different heights of the seedlings?* (4, 5, and 6 centimeters) Have students point to the scale that shows the heights and see the labels.
- **Say:** *The* ***Xs*** *show how many seedlings are each height. How can you find the most common height of the seedlings in the class?* (Look for the height that has the most **Xs**.) **Ask:** *Which height has the most* ***Xs****?* (6 centimeters) **Say:** *Most of the seedlings are 6 centimeters tall.*
- **Ask:** *How do you find the number of seedlings that are 4 centimeters tall?* (Look for the 4 on the scale and count the number of **Xs** above it.) *What number did you count?* (5)
- Assign students the appropriate activity page(s) to support their understanding of the skill.

Assess the Skill

Use the following problems to pre-/post-assess students' understanding of the skill.

Ask students to measure each other's height and create a tally chart. Then ask them to use the height data to create a line plot.

Name ______________________________________ **Date** __________

Use the data to complete the line plots.

1

Seedlings in Mr. Falber's Class	
Height in Centimeters	**Number**
4	IIII
5	𝍸
6	𝍸 II
7	II

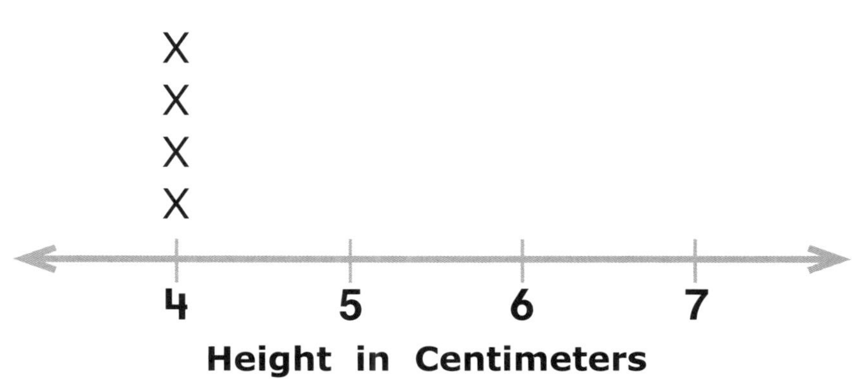

What is the least common height of the seedlings? ________

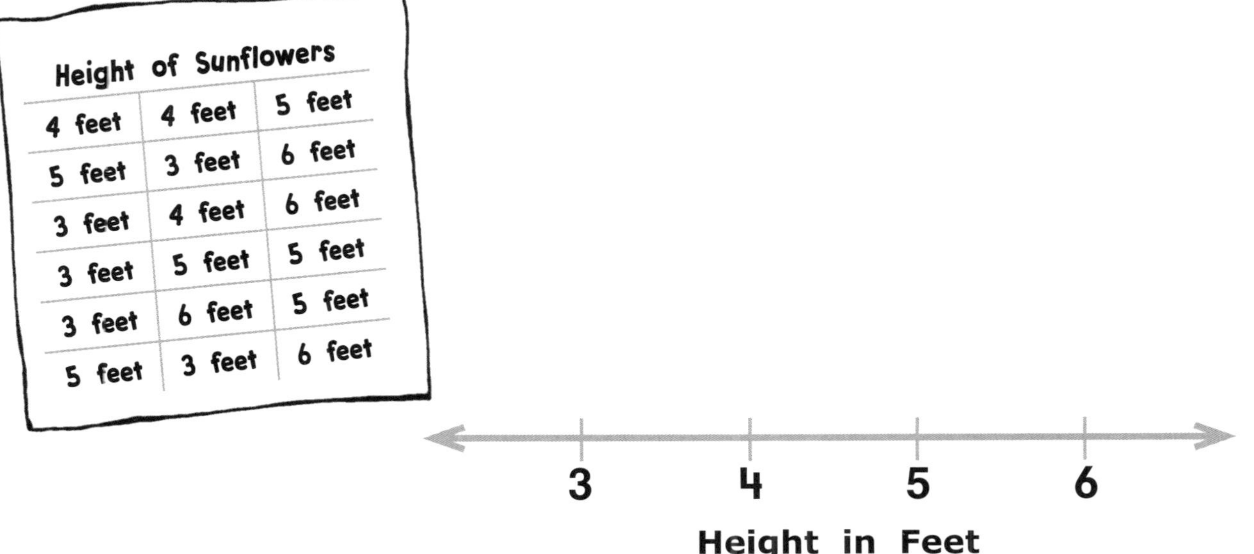

Height of Sunflowers		
4 feet	4 feet	5 feet
5 feet	3 feet	6 feet
3 feet	4 feet	6 feet
3 feet	5 feet	5 feet
3 feet	6 feet	5 feet
5 feet	3 feet	6 feet

How many sunflowers are more than 4 feet tall? ________

Tell how you completed a line plot.

Name ______________________________ **Date** __________

Use the data to complete the line plots.

Height of Bean Plants		
10 cm	9 cm	7 cm
8 cm	9 cm	9 cm
8 cm	9 cm	7 cm
9 cm	10 cm	8 cm
7 cm	8 cm	8 cm
10 cm	10 cm	9 cm

Height in Centimeters

What is the most common height for the bean plants? ________

2

Height of Cornstalks		
8 feet	7 feet,	5 feet
8 feet	8 feet	6 feet
8 feet	7 feet	6 feet
8 feet	7 feet	8 feet
6 feet	5 feet	6 feet
8 feet	6 feet	

____

What is the least common height for the cornstalks? ________

3

Height of Tomato Plants		
34 in	35 in	33 in
33 in	34 in	32 in
33 in	34 in	32 in
34 in	33 in	34 in
32 in	33 in	35 in
34 in	33 in	

____

How many tomato plants are less than 34 inches tall? ________

 **Tell how you know that you have plotted all the data.**

Name ________________________________ **Date** __________

Use the data to complete the line plot.

1. Make a line plot to show the length of the necklaces that Kylee made.

Length of Necklaces	
15 inches	17 inches
16 inches	18 inches
18 inches	16 inches
15 inches	18 inches
18 inches	15 inches

2. What is the most common length for the necklaces? ________

3. What is the least common length for the necklaces? ________

4. How many necklaces are more than 16 inches long? ________

   A 2
   B 3
   C 4
   D 5

5. How many necklaces did Kylee make in all? ________

   A 8
   B 9
   C 10
   D 12

 **Tell how you know what to label the scale.**

Unit 32 Mini-Lesson

Make a Graph

Standard

Data Analysis

2.10B (RS) Organize a collection of data with up to four categories using pictographs and bar graphs with intervals of one or more.

Model the Skill

Draw the following bar graph.

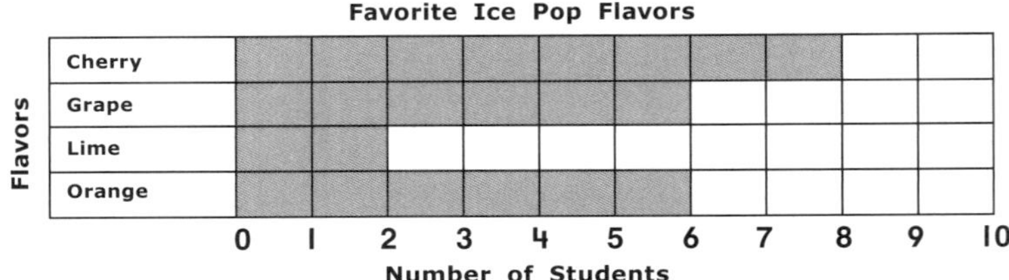

- **Say:** *Graphs can be used to show data. What does the bar graph show?* (Favorite Ice Pop Flavors) Have students look at the labels. **Say:** *One axis tells us the different flavors and the other shows a scale of numbers. What are the flavors?* (Cherry, Grape, Lime, and Orange) *To solve the first problem, find out how many students like cherry ice pops. Slide your finger on the bar labeled* ***Cherry*** *and then down to the number scale. The bar is at what number?* (8) Repeat the process for grape. (6) **Say:** *To find how many students like cherry and grape, add the numbers.* Write the number sentence. **Ask:** *How many students like cherry and grape ice pops?* (14)
- **Ask:** *What number are you going to subtract from 22 to find how many students like another flavor better than lime?* (the number of students who like lime: 2) Have students complete the number sentence to find the difference. (22 – 2 = 20)
- **Say:** *Now find the number of students that like orange and the number that like lime. Write a number sentence to compare the two amounts.* (Answers: 6; 2; possible equation: 6 – 2 = 4)
- Show the students the bar graph if it were drawn vertically.
- Assign students the appropriate activity page(s) to support their understanding of the skill.

Assess the Skill

Use the following problems to pre-/post-assess students' understanding of the skill.

Ask students to make a survey of the class shoe sizes and make a graph that shows shoe size data.

Name ______________________________ **Date** __________

Use the data to complete the bar graph. Then use the graph to solve the problems.

Favorite Subjects	
Subject	**Number**
Math	𝍸 𝍸
Reading	𝍸 III
Science	IIII
Social Studies	III

Favorite Subject

Subject										
Math										
Reading										
Science										
Social Studies										
	0	1	2	3	4	5	6	7	8	9	10

Number of Students

1. A total of 25 students are shown on the graph. How many students like another subject better than math?

____ students like math. 25 − ____ = ____

____ students like another subject better than math.

2. How many students like reading and science?

____ students like reading. ____ students like science.

____ ◯ ____ = ____

____ students like reading and science.

3. How many more students like math better than reading?

____ ◯ ____ = ____

____ more students like math better than reading.

4. How many more students like math better than social studies?

____ ◯ ____ = ____

____ more students like math better than social studies.

 **Tell how the graph helps you see the data better than a list or chart.**

Name ______________________________ Date __________

Use the data to complete the bar graphs. Then use the graphs to answer each question.

1

Favorite Sports	
Sport	**Number**
Soccer	卌 I
Baseball	II
Basketball	IIII
Football	卌 II

Favorite Sports

Soccer										
Baseball										
Basketball										
Football										
	0	1	2	3	4	5	6	7	8	9	10

Number of ______________

2 How many students like soccer better than basketball? ______

3 How many students like baseball and football? ______

4 How many students like a sport other than baseball? ______

5

Pets We Have	
Pet	**Number**
Dog	卌 卌
Cat	卌 III
Fish	II

____

Dog										
Cat										
Fish										
	0	1	2	3	4	5	6	7	8	9	10

____

6 How many more students have dogs than cats? ______

7 How many more students have dogs than fish? ______

8 How many more students have cats than fish? ______

☆ **Tell how you know what to label each axis of the graph.**

Name ____ Date __________

Use the data to complete the graph or pictogram.

Favorite Kinds of Games	
Games	**Number**
Card	𝍸 I
Board	III
Outdoor	𝍸 III

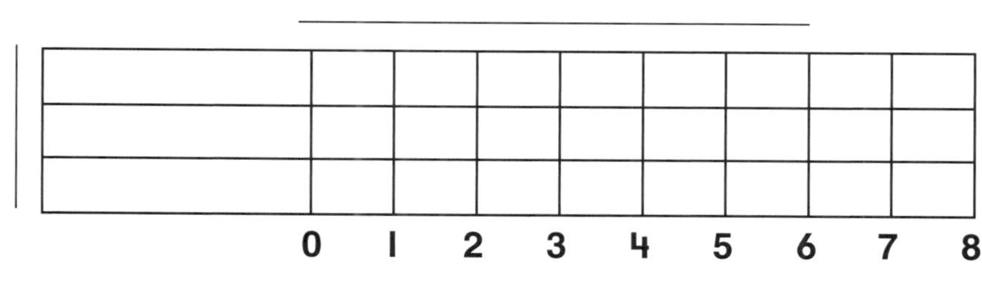

2

Favorite Colors	
Color	**Number**
Red	IIII
Blue	𝍸 II
Yellow	III
Green	𝍸 I

Favorite Colors

Red	☺ ☺ ☺ ☺
Blue	
Yellow	
Green	

Key ☺ = 1 student

3

Favorite Snacks	
Snack	**Number**
Pretzels	5
Yogurt	3
Fruit	8

____

______	
______	
______	

Key ________ = ________

Unit 33 Mini-Lesson

More Likely or Less Likely

Standard

Mathematical Process Standards

2.1F (PS) Analyze mathematical relationships to connect and communicate mathematical ideas.

Model the Skill

Use snap cubes in 2 colors and a paper lunch sack to demonstrate.

- Show students 4 white snap cubes and 1 red snap cube as you place them in the sack. Then draw squares on the board to represent the cubes.
- **Say:** *These squares represent the snap cubes in the lunch sack. Four squares are white and one square is red.*
- **Ask:** *If I pick a cube without looking, which color am I more likely to pick?* (white) *Why?* (more white than red cubes) *Which color am I less likely to pick?* (red) *Why?* Pick a cube from the sack and discuss the results.
- **Say:** *We can use the words* ***more likely*** *and* ***less likely*** *to describe things that are likely to happen. It is not certain that I will pick a white cube, because there is 1 red cube, but it is more likely that I will pick white, since there are more white cubes than red.*
- Continue the activity by putting a different number of snap cubes in two colors in the sack and have students describe the event of picking one color without looking as more or less likely.
- Assign students the appropriate activity page(s) to support their understanding of the skill.

Assess the Skill

Use the following problems to pre-/post-assess students' understanding of the skill.

Have students use the spinner to answer the questions.

Which shape is the spinner more likely to land on?

Which shape is the spinner less likely to land on?

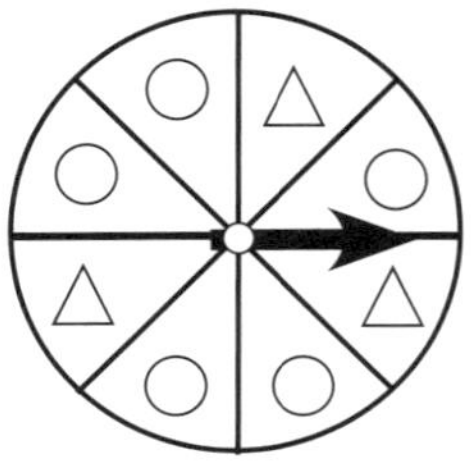

Name ____________________ Date ________

Look at the spinner for each problem. Then answer each question.

1

How many dogs? ________

How many cats? ________

Which pet is the spinner **more likely** to land on? ________

2

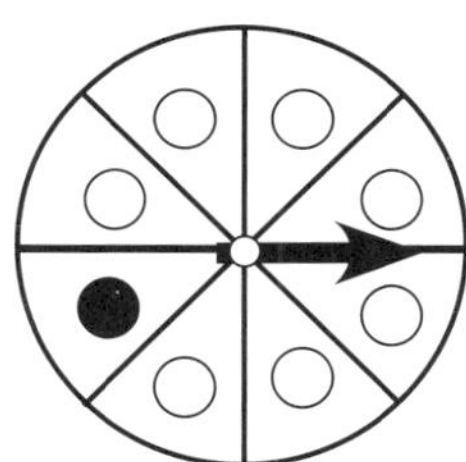

How many white circles? ________

How many black circles? ________

Which color circle is the spinner **less likely** to land on? ________

3

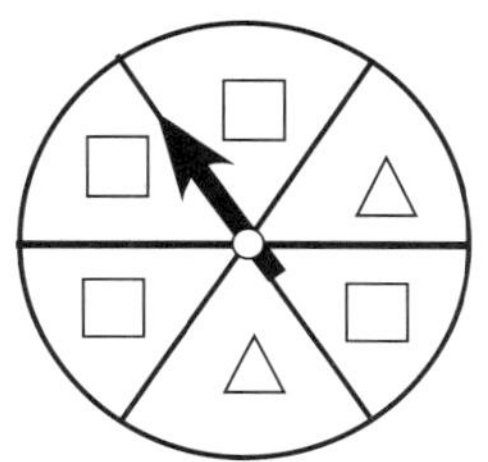

Are there more squares or triangles? ____________

Which shape is the spinner **more likely** to land on?

____

4

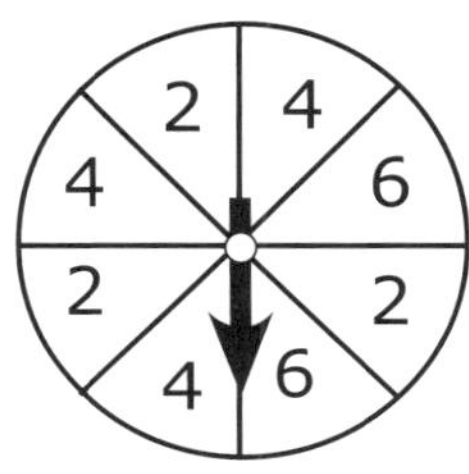

Are there fewer 2s, 4s, or 6s?

____

Which number is the spinner **less likely** to land on? ________

5

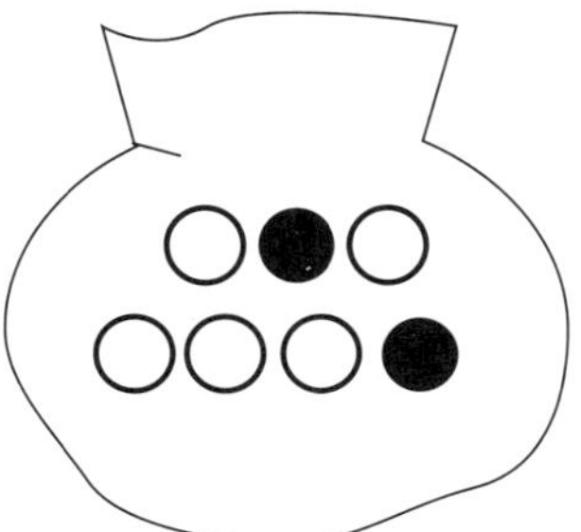

Ava picks a marble from the bag without looking. What color marble is she **less likely** to pick? ____________

6

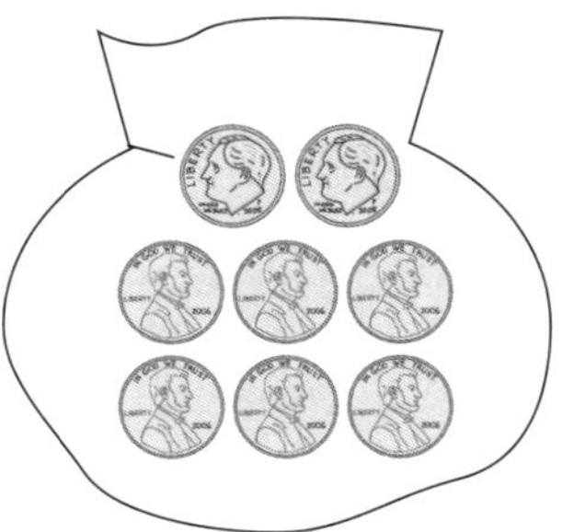

Jack picks a coin from the bag without looking. What coin is he **more likely** to pick?

____

☆ **Tell how you know which coin is more likely to be picked in Problem 6.**

Name ______________________________ **Date** __________

Draw a line to match each spinner to the sentence that best describes it.

1. The spinner is **more likely** to land on A.

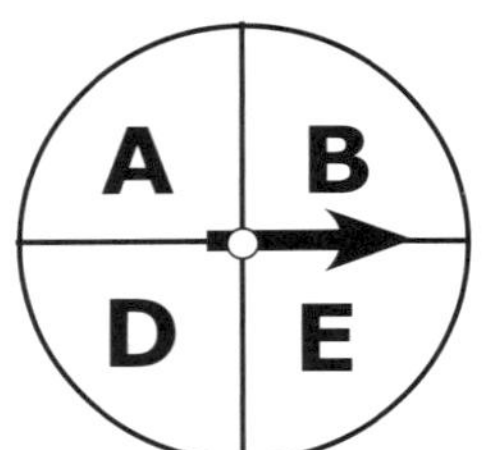

2. The spinner is **less likely** to land on B.

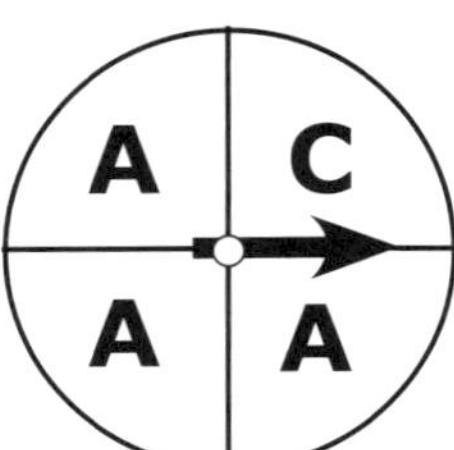

3. The spinner is **more likely** to land on B.

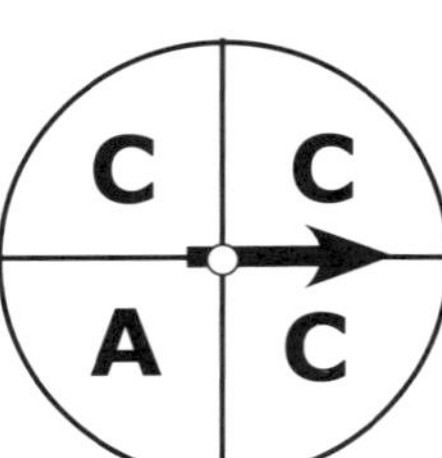

4. The spinner is **more likely** to land on C.

Answer each problem.

5.

Lana picks a coin from the bag without looking. What coin is she **less likely** to pick?

____

6.

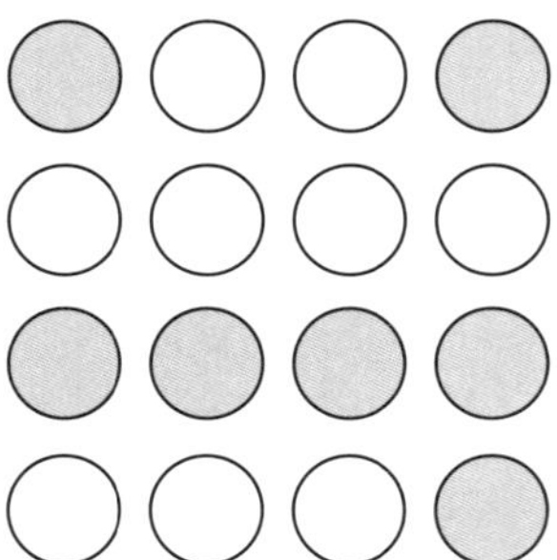

Chase picks a marble from the bag without looking. What color marble is he **more likely** to pick? ______________

 **Tell the steps you took to match spinners and sentences in Problems 1 and 2.**

Name ______________________________ **Date** __________

Solve.

Anna picked one of the marbles above without looking. Which color marble is she **more likely** to pick?

2

Luke spins the spinner. Which shape is he **less likely** to land on?

Circle the correct answer for each problem.

3 Jordyn spins the spinner. Which shape is she **more likely** to land on?

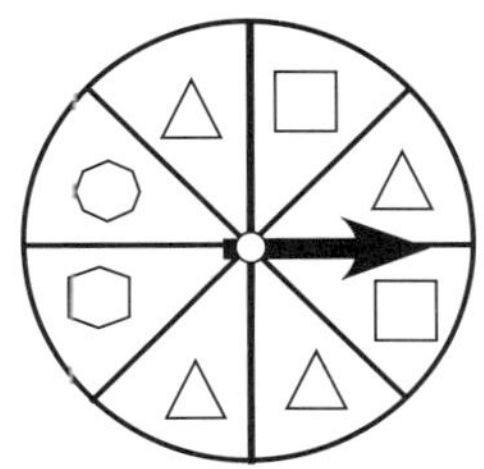

A squares

B triangles

C hexagons

D octagons

4 Rex picks a coin from a bag without looking. Which coin is he **more likely** to pick?

A penny

B nickel

C dime

D quarter

Unit 34 Mini-Lesson
How Much Money Is Saved?

Standard

Personal Finance Literacy

2.11A Calculate how money saved can accumulate into a larger amount over time.

Model the Skill

Show examples of several coins, including pennies, nickels, dimes, and quarters.

- **Ask:** *How much money do I have here?*
- **Ask:** *How can we save money over time?* (by putting it in a piggy bank)
- Assign students the appropriate practice page(s) to support their understanding of the skill.

Assess the Skill

Use the following problems to assess students' understanding of the skill.

If I put the following coins in a piggy bank, how much money will I save today?

____

Name ____ **Date** ________

If you save the following coins each day, write the amount of money you would save each day.

____

2

____

3

____

4

____

5

____

6

____

7

____

8

____

 **How do you find the total amount of money?**

Name ________________________________ **Date** __________

Inez made a table of the amount of money she saved each day for 4 weeks. Calculate the amount of money that Inez saved each week.

1

Week 1:	
Monday	25¢
Tuesday	30¢
Wednesday	40¢
Thursday	25¢
Friday	10¢

Saved: ____________________

Week 2:	
Monday	50¢
Tuesday	25¢
Wednesday	80¢
Thursday	20¢
Friday	10¢

Saved: ____________________

Week 3:	
Monday	$1.00
Tuesday	$1.25
Wednesday	30¢
Thursday	20¢
Friday	10¢

Saved: ____________________

Week 4:	
Monday	25¢
Tuesday	$1.00
Wednesday	$2.00
Thursday	$3.00
Friday	50¢

Saved: ____________________

2 What is the total amount that Inez saved for the four weeks?

____

Name ______________________________ **Date** __________

❶ Marcus has three coins that he had saved for his piggy bank on Monday, Tuesday, and Wednesday. The total amount of the three coins is 40¢. If Marcus saved a quarter on Monday and a nickel on Tuesday, which coin shows how much he saved on Wednesday.

❷ Sarah needs to save a total of $5.00 for the week. If the amount of money she saved during the week is shown in the table below, how much will Sarah need to save on Friday?

Monday	.25¢
Tuesday	$1.25
Wednesday	$1.25
Thursday	$1.25
Friday	?

❸ Discuss why you think it is important to save money.

❹ Saving money over time is a very important thing to do. Discuss different ways that you can save money over time.

Unit 35 Mini-Lesson
Deposits and Withdrawals

Standard

Personal Finance Literacy

2.11C (SS) Distinguish between a deposit and a withdrawal.

Model the Skill

On the board, write the words *deposit* and *withdrawal*.

- **Say:** *When you put money into your bank account, it is called a* ***deposit****.*
- **Ask:** *How many of you have seen your parents* ***deposit*** *money into a bank? What do you think happened to the money?*
- **Say:** *When you take money out of your bank account, it is called a* ***withdrawal****.*
- Assign students the appropriate practice page(s) to support their understanding of the skill

Assess the Skill

Use the following problems to assess students' understanding of the skill.

Do you think you have more or less money in your bank account after a deposit? ___________

Do you think you have more or less money in your bank account after a withdrawal? ___________

Name ______________________________ **Date** __________

Tell how much money you have in your bank account after each deposit or withdrawal.

1. Alex had $180 in the bank. Alex deposits $20. How much money does Alex have in the bank now?

2. Maria had $150 in the bank. Maria makes a withdrawal of $50. How much money does Maria have in the bank now?

3. Juan had $189 in the bank. Now Juan has $144 in the bank. Did Juan make a deposit or a withdrawal? Explain.

   By how much did Juan's amount of money change? Write a number sentence to solve the problem.

4. Shelby had $176 in the bank. Now Shelby has $210 in the bank. Did Shelby make a deposit or a withdrawal? Explain.

   By how much did Shelby's amount of money change? Write a number sentence to solve the problem.

☆ **What happens to your bank account if you make a withdrawal?**

Name ______________________________________ **Date** __________

Solve.

1. Max made a $40 deposit on Monday. Max deposited $50 more on Tuesday. On Wednesday, Max made a withdrawal of $25. How much money does Max have in the bank now?

2. Trace had $220 in the bank on Wednesday. On Thursday, he made a withdrawal of $100. On Friday, Trace deposited $50. How much does Trace have in the bank now?

3. Francis had $100 in the bank on Monday. On Tuesday and Wednesday, she made deposits of the same amount. Francis now has $160 in the bank. How much was each deposit?

4. Today Kelsey has $150 in the bank. Yesterday Kelsey made a deposit of $80. How much money did Kelsey have in the bank before the deposit?

☆ **What happens to your bank account if you make a deposit?**

Name ________________________________ **Date** __________

Solve.

1. Marilyn had $250 in her bank account at the beginning of the week. She made several deposits and withdrawals during the week. Complete the table to show how much money Marilyn had in her bank account after each deposit or withdrawal.

Day of the Week	Deposit or Withdrawal		Number Sentence Amount in Bank
Monday	Deposited	$20	$250 + $20 =
Tuesday	Withdrawal	$50	
Wednesday	Deposited	$25	
Thursday	Deposited	$75	
Friday	Withdrawal	$80	

2. Write a problem that would show a deposit.

3. Write a problem that would show a withdrawal.

Notes: